D1068321

WITHDRAWN

THE GEOGRAPHY OF SCIENCE

HAROLD DORN

=

THE GEOGRAPHY OF SCIENCE

THE JOHNS HOPKINS UNIVERSITY PRESS

BALTIMORE AND LONDON

The Johns Hopkins University Press
701 West 40th Street, Baltimore, Maryland 21211
The Johns Hopkins Press Ltd., London

The paper used in this book meets the minimum
requirements of American National Standards for
Information Sciences—Permanence of Paper for
Printed Library Materials, ANSI Z39.48-1984.

*Library of Congress
Cataloging-in-Publication Data*
Dorn, Harold, 1928–
The geography of science / Harold Dorn.
p. cm.
Includes bibliographical references and index.
ISBN 0-8018-4151-8 (alk. paper)
1. Science—history. I. Title.
Q125.D65 1991
509—dc20   90-25149

In memoriam
GEORGE AND BECKY DORN
who worked with their hands

History celebrates the battlefields
wherein we meet our death, but it
scorns to speak of the plowed fields
whereby we thrive. It knows the
names of the King's bastards, but it
cannot tell us the origin of wheat.

Jean Henri Fabre

CONTENTS

===

PREFACE

This book originated in academic banter. I jestingly proposed to teach a course on science in the Asiatic Mode of Production, whereupon a light seemed to flash and loose bits of information instantly rearranged themselves. The sparkle of recognition illuminated a strange landscape where science appeared in a terrain diversified, not by philosophical and intellectual features, but instead by technological, ecological, and geographical conditions.

Over the next few weeks the light dimmed as its energy was transformed into work. The initial clarity was beclouded by a mist of facts and qualifications. The exuberant sense, in Albert Szent-Gyorgyi's expression, of "having looked at what everyone has looked at and seen what no one has seen" soon gave way to uncertainties about the validity of a history of science embedded in a matrix of geographical explanation. Nonetheless, there remained a residual sense that *something* had been seen in that flash of enlightenment. So I pressed my search and soon found myself trawling in the exotic waters of anthropology, archaeology, prehistory, and geography.

When science is situated and studied in its physical environment, its development appears nearer to geographical considerations than historians have generally placed it—to conditions of soil, climate, hydrology, and topographical relief, and to demographic fluctuations, latitude, and the differences between sown fields, steppe, and desert. From that angle of view the history of science is seen against a background of the material settings in which societies take root, flourish, and decay. It is akin to the histories of agriculture, technology, and industry, and of the institutions that regulate not only the development of these activities but also the conduct of scientific research and the practice of statecraft.

Insofar as this book establishes the plausibility of the subject, it can only be a prolegomenon to further inquiry and analysis. The components of a geography of science are available in numerous studies of ecological and geographical variables, of the processes of scientific change, and of the history of institutions. A survey of these components, along with reflection on how they might be assembled, is all that the present work pretends to be. If it succeeds in eliciting discussion and inducing a few readers to study the interconnections between scientific culture and material life more carefully and critically, the main objective of the book will have been fulfilled.

Despite my somewhat reclusive work habits, the list of persons to whom I owe acknowledgment is long. It is possible to cite here only a few individuals. My friend and colleague, Professor James E. McClellan III, of the Stevens Institute of Technology, who goaded me to the limits, and beyond, of sociohistorical conjecture, read and criticized drafts of each chapter. His searching questions combed out a few serious mistakes and his invariably helpful comments contributed a few valuable suggestions. Professor Miriam Selchen Dorn, of the City College of New York, read the penultimate draft and removed some of the excesses of a style that, over the years, she has frequently kept within bounds. The logic and language were helpfully criticized by Professor Ricardo Otheguy, also of CCNY. Many enlightening conversations with Professor Theodore Brown, University of Rochester, over the years of composition clarified the major thesis. Syl McNinch, at the National Science Foundation, supplied me with information that could not have been found in any published source. The manuscript was also fortunate in receiving the exacting criticism of Professor Charles Gillispie, Princeton University. And I have the pleasant obligation of citing him again for having always led his students in the history of science away from every cramped parochialism and towards an expansive conception of the subject. The extent to which I have carried his encouragement too far is my own fault.

A few impersonal acknowledgments are also in order. I have been newly enlightened on an old subject by G. A. Cohen's discourse on Marx's analysis of history and the structure of society (G. A. Cohen, 1978). Ellen Churchill Semple's learned observations on the geography and culture of ancient lands remain fresh after sixty years (Semple, 1931). Edward Hyams' reverence for the soil has been an inspiration and should, incidentally, reassure those who worry about an inhumane "geographical determinism" (Hyams, 1976). Joseph Needham's vast mindscape depicting the geography and the cultural history of China is endlessly suggestive of how the two subjects can be connected in other settings as well (Needham, 1954–). And the *Dictionary of Scientific Biography*, edited by Charles Gillispie, is that rare reference work that can be read and not merely consulted, and that gives pleasure as it instructs (Gillispie, 1970–1980).

Biographers are known to develop sympathy and sometimes affection for their subjects, and concern over their character defects. Social and political historians, even while denying Whiggish prejudices, often show a preference for constitutional regularities and a distaste for invidious distributions of social benefits or political rights. In like manner, searching

for the geographical and technological components of historical understanding has its own biases—an affection for maps and atlases, for descriptions and drawings of plows, harnesses, shadoofs, sluice gates, and
qanats; admiration for those who adapted so ingeniously and indus-
triously to difficult physical conditions; and respect for the power of
technique and the illumination of knowledge. But it also stirs a deepen-
ing concern over the despoliation of the Earth as a habitat, and it evokes
an ethic of conservation and replenishment. In the ancient world, immoderate agricultural intensification left a ruined estate of dead and
buried cities, layer upon layer. In the modern world, rapacious agricultural and industrial intensification have raised the specter of a spent
and stricken planet. To bequeath a spoiled landscape to those who come
next would be the ultimate cataclysm. The scale of the calamity is global,
and the remedy, which is not yet even discernible, must surely be global
too. But if past is prologue, remedies will be found and agreed upon only
when the destruction is universal, leaving no sanctuaries where, like
Carthage and Nineveh and Memphis, power and privilege could detach
themselves in splendid isolation and continue to see only goodness and
beauty—until it was too late. The processes of balance and renewal and
continuity, which early agricultural man, before the decay set in, took to
be certainties and recorded in his books of wisdom, have become disturb-
ingly equivocal. Will the Earth remain intact as a secure habitat for the
improvement of technique and the extension of knowledge? Can we still
be confident that

> One generation passeth away,
> And another generation cometh;
> But the Earth abideth forever.

INTRODUCTION

===

There is a famous photograph, taken at the 1927 Solvay Congress in Brussels, of an assemblage of twenty-nine of the world's leading scientists. All of them were Europeans and almost all of them worked in Western Europe. If a comparable photograph could have been made in the tenth or eleventh century most of the scientists would have been Muslims, or at any rate would have lived and worked under Islamic auspices in Spain, North Africa, the Middle East, and Central Asia. In the third century B.C. the group portrait would have shown individuals who did their research in Alexandria and several other cities of the central and eastern Mediterranean. And going back another thousand years it would have portrayed the scribes who devised calendars and invented mathematics in Egypt and Mesopotamia.

Ever since their origins in the ancient Middle East, scientific cultures have displayed geographically uneven patterns of growth, retardation, and decline or have been totally absent. Such geographical variations in the scope, quality, and intensity of scientific activity have appeared not only over stretches of time; they can also be seen within any given period. In the seventeenth century, when Galileo, Descartes, Newton, Huyghens, Leibniz, Hooke, and Boyle, working in Italy and Northern Europe, were revising the content of the old sciences while creating new ones, hardly any scientific research was being conducted in the Americas. Never has it been carried on in nomadic societies anywhere.

These geographical observations have remained barely visible in systematic studies of scientific development. It is generally assumed that scientific research displays historical, philosophical, and sociological patterns—it changes over time with intellectual currents and in step with social, political, and institutional developments. While these are indeed often the primary determinants of differences in scientific cultures, they are not the only ones. Science also changes over space and in the context of environmental conditions, and in some situations its development has as much to do with geography as with history.

Correlating historical processes with geographical conditions is not widely practiced as an academic vocation. But neither has it been totally neglected. In the history of science its most eminent practitioner is the

British polymath Joseph Needham. Thirty-seven years ago he published Volume One of what is generally regarded as a towering achievement of twentieth-century scholarship—*Science and Civilisation in China* (1954–). In that first volume Needham raised some of the theoretical issues that the geography of science inevitably encounters. He began with a "geographical introduction," and in the plan of the work he indicated that the final volume would be devoted to the question, "Why, therefore, did *modern* science, the tradition of Galileo, Harvey, Vesalius, Gesner, Newton . . . develop round the shores of the Mediterranean and the Atlantic, and not in China or any other part of Asia?" The answer that Needham projected involved "an examination of the concrete environmental factors of geography, hydrology, and the social and economic system which was conditioned by them." And he suggested that a full analysis would examine and interconnect intellectual, sociological, geographical, ecological, and technological factors—the "extent to which the forms of Chinese society were determined by the hydrology of its environment; the constant need for works of hydraulic engineering (irrigation, flood protection, tax-grain transportation)" (Needham, 1954: 19, xxxviii).

The methods of the geography of science are manifold and diverse. One approach to the subject ventures into theoretical sociology. Needham's geographical and ecological interpretation of the history of Chinese science and society is an example of that method. It springs from a few suggestive ideas enunciated by Marx and Engels (and consistently repudiated by most Marxists ever since) to the effect that the traditional Oriental societies, practicing what Marx designated as the "Asiatic Mode of Production," developed along the arid belt of the Old World, from Egypt, through Mesopotamia and India, to China. In order to conduct agriculture, which formed the basis of the economy in these regions, centralized state authorities came into being to control vital hydrological resources and techniques, including artificial irrigation. These same authorities patronized the science of astronomy insofar as they judged it to be useful in establishing the rhythms of the agricultural seasons and, in the form of astrology, in predicting the future.

Chapter 1 presents the genealogy of these ideas, often referred to in their non-Marxist expression as the "hydraulic hypothesis," and suggests how a binary model of scientific development in the ancient world might be based on them. At one pole of the model is the bureaucratic organization of state-sponsored research in politically centralized societies where the central authority is responsible for the hydraulic engineering projects necessary in arid and semi-arid regions. At the other pole is the individual savant passionately engaged in pure research in societies where the basic economy calls for little intervention by the central government. Historically, these poles are represented respectively by the Bronze Age kingdoms in Egypt and Mesopotamia and by Hellenic Greece. In Chapters 2 and 3 the bipolar model is examined in greater detail by contrasting four "Asiatic" scientific cultures with Hellenic Greek science viewed in its social context.

Chapter 4 describes the orientalization of Greek science in the strongholds of the ancient Middle East as Greek culture followed in the wake of Alexander's conquests. Along with this Hellenistic stage of Greek science, centered in Alexandria, three additional Middle Eastern civilizations that flourished in late antiquity and the medieval era are discussed—the Byzantine Empire, Sassanid Persia, and Islam. Each of these combined in some measure characteristics of a Hellenic intellectual tradition and an Oriental pattern of state-supported cultural institutions.

The geography of science, as developed in this book, generally directs attention to the material circumstances that have tempted the state to intervene in the basic economy and in the organization of scientific research. In Northern Europe, the development of the state's role in the social organization of science followed a deviant path. Chapter 5 examines the rise of modern science in northwestern Europe in the context of the unique environmental conditions of the European landscape. An agricultural revolution that expanded the economic base, the rise of cities and the European university, the Black Death that depressed scientific activity for more than a century, and the "gunpowder revolution" that triggered the centralization of state authority were all part of the process. After the fifteenth century, centralized nation-states came into being north of the fortieth parallel of latitude and began to patronize a steadily rising level of scientific activity and achievement.

In Chapter 6 the hydraulic hypothesis is again applied, and its argument is indeed fortified by the surprising case of science among the Utah Mormons.

Historians and geographers have adopted overlapping but dissimilar attitudes towards nature-culture studies. When historians bring environmental conditions into focus, it is specifically to measure their effects on social and historical processes.[1] For the geographer, nature-culture studies are one dimension of a general interest in space and place.[2] The historian's emphasis is more likely to provoke methodological and philosophical objections, and it is perhaps incumbent on the geographical interpretation of the history of science to anticipate and confront these objections. In the nineteenth century, the preeminent student of cultural geography, indeed the "true founder and greatest single contributor to the development of modern human geography," was Friedrich Ratzel, a German geographer who produced a voluminous corpus of writing, largely on political and anthropological geography (Beckinsale, 1975: 309). The "anthropogeography" that Ratzel advocated inevitably directed attention to differences in physical environments and their populations, as well as to the politically uncomfortable fact that resources and material conditions are unevenly distributed. Despite his disclaimers, some of his work appeared to draw invidious distinctions among ethnic groups; and his writings on space as a political concept—*lebensraum*—acquired an ominous accent in 1933, so much so that his contributions have been consigned, at least by historians, to a secular *Index Librorum Prohibitorum*.[3]

The connections between geography and history were further complicated by the American geographer, Ellsworth Huntington, "the most notable exponent of environmentalism in the English-speaking world in the twentieth century" (Spate, 1968: 26). What was discernible in Ratzel was only too clear in Huntington. Among environmental agents, he singled out climate, and he correlated the climate of higher latitudes with "the superiority of the white race" (Huntington, 1922: ch. 2). Huntington had many more interesting and important things to say, but his "new science of geography"—comparing "the physical and organic maps and thus determining how far vital phenomena depend upon geographic environment"—buckled under the weight of his exaggerated monocausality and his racial conjectures. What survived instead, in large measure, was what he called the "old geography"—producing "exact maps of the physical features of the earth's surface." The combination of a hard determinism and a baseless raciology would have been enough to discredit the hundred great books. Its effects on the small field of human geography were proportionately more devastating.

Fortunately, there are signs that cultural geography and environmental history can be acquitted of the charges that have tainted their reputations, and it seems clear that they can be developed as sciences without an ideological agenda. Ratzel's American disciple, the geographer Ellen Churchill Semple, presented an interpretation of his anthropological geography to English-speaking readers and applied some of his ideas in her own irreproachable studies of American history and the history of the ancient Mediterranean world (Semple, 1911; 1931).[4] In French scholarship there is a long historiographical tradition in which close contact has been sustained between geography and history. Among its more recent expressions is Emmanuel Le Roy Ladurie's study of the history of climate in Europe (Le Roy Ladurie, 1971); and the eminent French scholar Fernand Braudel was as much a geographer as he was a historian. In the United States the combination of history and geography has recently been served well by a number of historical studies that gives signs of developing into a groundswell: William McNeill's study of the historical effects of medical geography, *Plagues and Peoples* (1976); Donald Worster's account of the political implications of the geography of American agriculture, *Rivers of Empire* (1985); his *Nature's Economy: A History of Ecological Ideas* (1977); and his edited collection, *The Ends of the Earth:*

Perspectives on Modern Environmental History (1988); D. W. Meinig's survey of American history in geographical perspective, *The Shaping of America* (1986); and Alfred Crosby's geography and macro-history of European expansion across the world, *Ecological Imperialism* (1986). In addition, two recent anthologies indicate the scope of the developing interest in the integration of history and geography: Kendall E. Bailes, ed., *Environmental History: Critical Issues in Comparative Perspective* (1985), and Eugene Genovese and Leonard Hochberg, eds., *Geographic Perspectives in History* (1989).

In 1979 the history of climate and the possibility of integrating it with general history—"research on the impact of climate on man's works and events"—were promoted by a conference at Princeton University sponsored by the *Journal of Interdisciplinary History* (see Rotberg & Rabb, 1981). In a summary presentation, the historian David Hackett Fischer stated: "Since Huntington's day, historians and social scientists have generally turned away from questions of climatological determinism. But the Huntington thesis was never really refuted. It was merely ridiculed, because it failed to fit the metaphysical framework of social science in the mid-twentieth century. Perhaps it is time for those issues to be reopened" (Fischer, 1981: 248).

It may now be timely not only to reconsider the general connections between history and geography but also to examine specific cultural developments in their geographical and ecological settings. In the context of the history of science, Needham's studies have thrown wide open many issues in the relationship of geography and culture less confined than the specific question of "climatological determinism." In dialectical opposition to Huntington's ethnic biases, Needham came to his work under the banner of a Marxist evangelical ecumenism. In his world-view, Marxist convictions applied to the social history of science lead to the most fraternal conclusions:

Perhaps this is where the Marxists (somewhat to their surprise, no doubt) contribute to a juster humility of Europeans. For if it was largely external social and economic circumstances that brought about the scientific revolution, then the intellectual capacities of Westerners may not be (as so many of them have believed) that much superior to those of other peoples. The sheer intellectual power of the giants of the Newtonian tradition might blind many (as it has tended to do in the past) to the achievements of other traditions, and perhaps if the social and economic circumstances had been right in other cultures, modern science would have arisen somewhere else than where it actually did. (Needham, 1973: 6)

And on the historian of science as "ecumenical man":

Seeking first the Kingdom of God today means accepting all men everywhere on a basis of absolute equality and fraternity, and seeking justice everywhere for all human needs. (Needham, 1973: 8)

Needham's fraternalism is clearly more appealing than Huntington's bogus physical anthropology. But neither of them carries any weight in social analysis. Huntington's ethnic prejudices have no place in the social sciences; and Needham's study of Chinese science must stand on its argument, not its spiritual edification. The history and the geography of science have the power to enlighten us, but they cannot teach us to be good. That lesson is learned elsewhere. Scholars in the social sciences can study the world, but since their studies can neither change it nor prevent it from changing, they cannot be blamed for it.

===

CHAPTER ONE

SCIENCE AND MATERIAL LIFE

FROM KARL MARX TO KARL WITTFOGEL

During the 1850s, in the course of his employment as the London correspondent of the *New York Daily Tribune*, Karl Marx turned his attention to British rule in India. He had never closely studied a non-European society and he now confronted fundamental theoretical questions about the civilizations of the East. What essential economic conditions underpinned those societies? What distinguished them from Western civilization? And what accounted for their scandalous resistance to change despite their sharp and appropriately dialectical antagonisms? Hegel had perceived a formidable dialectic to which he attributed the arrested development of these ancient social systems, and he described it in terms of a poetic metaphor. The sphinx, man emerging from beast, symbolizes the history of the Eastern civilizations, the long, painful historical epoch during which society ("spirit") began to break the bonds of nature:

Of the representations which Egyptian antiquity presents us with, one figure must be especially noticed, viz. the Sphinx—in itself a riddle—an ambiguous form, half brute, half human. The Sphinx may be regarded as a symbol of the Egyptian spirit. The human head, looking out from the brute body, exhibits spirit as it begins to emerge from the merely natural—to tear itself loose therefrom and already to look more freely around it; without, however, entirely freeing itself from the fetters nature had imposed. (Quoted in G. A. Cohen, 1978: 11)

Marx, however, was seeking the reality behind the symbol, rather than its Hegelian opposite. For him, the key to understanding Oriental societies was to be discovered in their economic infrastructures, not in their mythic sculpture. On June 2, 1853, writing from London to Engels in Manchester, he declared: "Bernier rightly considered the basis of all phenomena in the East . . . to be the *absence of private property in land.* This is the real key, even to the Oriental heaven" (Marx & Engels, 1980: 277). Four days later, Engels replied with characteristic deference, "The absence of property in land is indeed the key to the whole of the East." Then, having paid his respects, he probed more deeply than Marx and

1

reached an acute insight. Searching for the material substratum from which these property relations arose, Engels framed the geographical interpretation of the river-valley civilizations of the East:

But how does it come about that the Orientals did not arrive at landed property . . . ? I think it is mainly due to the climate, taken in connection with the nature of the soil, especially with the great stretches of desert which extend from the Sahara straight across Arabia, Persia, India and Tartary up to the highest Asiatic plateau. Artificial irrigation is here the first condition of agriculture and this is a matter either for the communes, the provinces or the central government. (Marx & Engels, 1980: 278)

Not all writers have looked favorably on Engels' attempt to buttress Marx's sociological conjecture with a physical explanation. Some schools of Marxist thought, determined to drain it of its empirical content and preserve it as a philosophical relic, have preferred a more dialectical logic, whereby cause and effect are so safely interpenetrated that no empirical evidence can ever discredit a favored opinion. For his clear, if arguable, geographical explanation of the structure of Oriental societies Engels has been reproached for regressing to the Enlightenment tradition of "ordinary" thinking:

Engels's notion that the failure of Oriental society to develop private landed ownership was "mainly due to the climate" is a trifle naïve, and looks back to Hegel or even Montesquieu: one of the many instances of his tendency to relapse into ordinary cause-and-effect explanation, in the manner of the Enlightenment. (Lichtheim, 1963: 94n)

In his later writings Marx did not attach much weight to geographical variables, possibly because he was for the most part studying the transition from feudalism to capitalism in a single geographical setting— Northern Europe. And with regard to the specifics of technological factors like irrigation engineering, Marx exaggerated the importance of property rights and the ownership of wealth and underestimated the explanatory role of tools and technique. His often-quoted aphorism, "the hand-mill gives you society with the feudal lord; the steam-mill, society with the industrial capitalist," might have become theoretically fundamental to his sociohistorical analysis; but he never systematically developed its implications in the empirical context of different social systems erected on different technological infrastructures. Instead, Marx generally chose, as in his tribute to Bernier, to place his analytical emphasis on the economic level ("the absence of private property in land") rather than on the technological or, even less, on the ecological level. He categorized societies as social formations, not ecological systems. To do otherwise, to displace the prime mover of social analysis and change from property relations to a geographical substratum, as Engels had suggested, would have appeared to be replacing his program of political action with its opposite, a program of observation and study of impersonal physical forces which political action cannot change. Nonetheless, in 1853 Marx was sufficiently persuaded by Engels' argument to repeat it (nearly verbatim but without attribution) in his column in the New York Daily Tribune (Marx & Engels, 1980: 33).[1]

In addition to the sequence of aridity to irrigation agriculture then to communal property and centralized authority that Engels postulated, his seminal suggestion possesses further analytical possibilities. The social organization and political centralization associated with the control of water for irrigation, the redistribution of increased food surpluses, and the large-scale engineering projects required for agriculture in arid and semi-arid regions might jointly define a hydraulic society whose basic economy was largely determined by ecological conditions. The construction and maintenance of systems to provide irrigation water and prevent destructive inundations would necessitate an army of laborers—Lewis Mumford's "megamachine"—available on demand and equipped and sustained at great cost; an engineering bureau would be required; irrigation water would need to be allocated, and disputes between upstream and downstream contenders and rural and urban needs would call for expeditious and sometimes peremptory settlements; and the surplus grain produced by intensive agriculture would need to be taxed, stored, guarded, and redistributed.

All of these features of a hydraulic society might bring forth a despot, a water king, who would maintain the seasonal rhythms and long-term continuity of irrigation agriculture. Moreover, should the despot fall, his conqueror, whatever enlightened views he might hold, would soon find that his first, most compelling order of business was the supervision of a labor corvée to attend to the dredging of canals clogged with silt, the repair of inoperative sluices, the filling of breached embankments, and the construction of new tanks and catchments. The conqueror would be crowned, a new despot would ascend the throne, a new dynasty would be generated, and the immutable East would close ranks against change while the millennia slipped by. Herein might lie the explanation, not only of "Oriental despotism," but also of the distinctly non-Marxist inertia of the ancient Oriental civilizations, characterized by revolutions without reform, history without stages, dialectics without synthesis.[2] Even science was not beyond the reach of the geographical principle. Might not the despot, responsible for the operation and security of a regionwide system of irrigation agriculture, require the official services of "calendar-Brahmins" and learned prognosticators?[3]

Over the second half of the nineteenth century Marx and Engels intermittently touched upon each of these analytical possibilities, albeit with sufficient inconsistency and ambiguity to provoke a century of heartfelt controversy among zealous idealogues over what was said and what was meant.

In 1859, Marx confronted the issue of social classification—a taxonomy of societies—and the "material conditions" that gave rise to them. He recognized six types of societies, which he treated as historical stages forming a sequence: four "progressive epochs," marked by social antagonism, and two nonantagonistic stages (primitive communism at the beginning and mature communism in the future). And he associated these forms of society with characteristic modes of production: "In broad outlines Asiatic, ancient, feudal and modern bourgeois modes of production can be designated as progressive epochs in the economic formation of society" (Marx, 1977: 390). He thus specified an "Asiatic mode of production" (asiatische Produktionsweise), corresponding to what is generally termed the Bronze Age or the ancient Oriental civilizations, as one of his six major social types.

In 1878, Engels proposed that the state arose from and perpetuated itself around the need to satisfy supreme social functions that, in the arid East, were mainly connected with the requirements of irrigation agriculture:

Here we are only concerned with establishing the fact that the exercise of a social function was everywhere the basis of political supremacy; and further that political supremacy has existed for any length of time only when it fulfilled its social functions. However great the number of despotic governments which rose and fell in India and Persia, each was fully aware that its first duty was the general maintenance of irrigation throughout the valleys, without which no agriculture was possible. (F. Engels, 1939: 198–99)[4]

Marx, too, occasionally noted the sociological importance of large-scale public works. He stated that in the East "the public works were the business of the central government" and he considered the inescapability of that duty to be one of the causes of the "stationary character" of Indian society (Marx & Engels, 1980: 279). On science, Marx stated in Das Kapital the opinion that calendrical astronomy, "and with it the rule of the priest caste as leader of agriculture," followed from the need to predict the action of the Nile (Marx, 1936: 564).[5] Like most nineteenth-century writers who valued science for its contribution to positive knowledge, Marx called attention to calendrical astronomy only to the extent that it provided a rational guide to the practice of agriculture. Since then, however, it has become amply clear that ancient science (and even modern science until the eighteenth century) drew few distinctions between rational and occult knowledge. Along with calendar-keeping in the interests of agriculture, the royal patrons in the ancient East, and hence the savants they employed, were at least equally interested in the reputed practical applications of higher learning in the fields of astrology and magic.

This entire constellation of opinions, encompassing the multiple relationships among ecology, society, and science, has been held in various combinations by many writers, and it is by no means a hallmark of either Marxism or materialism. The term *Oriental despotism*, which Marx adopted, can be traced back to the seventeenth century. In the eighteenth century Adam Smith commented on the centralization of authority necessitated by hydraulic engineering in the ancient kingdoms—"works constructed by the ancient sovereigns of Egypt for the proper distribution of the water of the Nile" (Smith, 1937: 646). John Stuart Mill drew a distinction between the agricultural communities of early Europe and the "modes of appropriation" of "Oriental nations," with their "tanks, wells, and canals for irrigation, without which in many tropical climates cultivation could hardly be carried on" (Mill, 1961: 12–13). Max Weber, referring to the ancient Near East, remarked that "the basis of the economy was irrigation, for this was the crucial factor in all exploitation of land resources. . . . Here then is the fundamental economic cause for the overwhelmingly dominant position of the monarchy in Mesopotamia (and also in Egypt)" (Weber, 1976: 84). And many others, including numerous American anthropologists, have taken up the possible connections between the ecological constraints on agriculture under severe hydrological conditions (mainly insufficient moisture) and the organization of society.[6]

The geographical interpretation of the ancient Oriental civilizations has provoked sharp polemics on both sides of the ideological divide. Outside of Marxist circles it is often referred to as the "hydraulic hypothesis," in lieu of Marx's faintly irritating designation, the "Asiatic Mode of Production" (Marx was surely innocent of any racial insult lurking in his expression); and the argument has turned mainly on the pivotal importance attached to irrigation and to water-control projects in general as determinants of social and political organization. On a theoretical level, accusations of monocausality and reductionism, of reducing history to geography, have been leveled against proponents of the hypothesis.[7] And on an empirical level, the claim that bureaucratic elites, political centralization, or despotism has been the result of large-scale water control projects has been challenged in the context of specific hydraulic societies.

In Marxist and Soviet circles, the concept excited a fiercer and darker dispute, spilling over into the arenas of practical politics and privileged opinion. Some communicants who have found the geographical interpretation objectionable have taken refuge among Marx's and Engels' equivocations on the issues. Chinese communism complicated the situation further; and during the 1920s Mao Tse-tung and his colleagues recognized no Asiatic Mode of Production as a stage of historical development, citing only "classless primitive communes," slavery, and feudalism prior to the nineteenth century (Mao Tse-tung, 1954: 73–75). The problem was finally judged to be too momentous to be safely entrusted to rational discourse. In February 1931, at a fateful conference in Leningrad, the issues were debated, and the theory of the Asiatic Mode of Production was sharply condemned on openly ideological grounds. The quality of the criticism can be judged from the remarks of one of the discussants:

> The theory of the Asiatic mode of production could be used not only for the purposes of studying agriculture in the Chou dynasty, but has been used, and can still be used, to dispute the pronouncements of the Communist International on questions of the nature of colonial revolution. In this respect the concept of the Asiatic mode of production is the wet nurse for the theoretical position of Trotskyism. . . . A theory which is useless in application to the present day must be discarded. (Godes, 1981: 104)

Although some of the discussion that took place at the conference was conducted along more rational and analytical lines, the concept was repudiated; and for the next thirty years the idea was ignored by Marxist writers, with few exceptions. Then, during the 1960s, the Asiatic Mode of Production was rehabilitated along with other casualties of the 1930s, and the rehabilitation produced a new debate that resulted in numerous publications inside and outside of the communist world.[8]

One of the few Marxists who considered the role of geographical factors in historical explanation after rejection of the Asiatic Mode of Production was Stalin. In 1939, he reviewed the geographical interpretation of history and decreed it to be in error. "There can be no doubt," he wrote, "that the concept 'conditions of material life of society' includes, first of all, nature which surrounds society, geographical environment, which is one of the indispensable and constant conditions of material life of society and which, of course, influences the development of society. What role does geographical environment play in the development of society?" He answered his own question by stating that while geography "accelerates or retards [society's] development," it could not be the "determining influence" on society, since the pace of geographical (by which he seems to have meant geological) change is too slow to account for historical events. Then, without mentioning the concept, he wrote the Asiatic Mode of Production out of the panoply of the Marxist theory of history. Where Marx had specified six major types of society there were now five: "Five main types of relations of production are known to history: primitive communal, slave, feudal, capitalist and Socialist" (Stalin, 1939: 118, 123).[9]

In the renunciation of the Asiatic Mode of Production, the whole set of issues surrounding the interconnections between the technical and the administrative requirements of hydraulic agriculture in arid and semi-arid regions, on the one hand, and the patterns of social organization, on the other, had been abandoned—but not utterly. The communist movement, in truly dialectical fashion, is renowned not only for discipline but also for apostasy. During the 1920s, Karl Wittfogel was a member of the German Communist Party; later, as a political refugee, he found himself in China, where he embarked on a study of Chinese history in which he stressed the importance of irrigation agriculture in ancient China and referred to the "Asiatic system of production." Then, after the German-Soviet nonaggression treaty (1939), he renounced communism, and with the apostate's grim determination to injure the object of his former adoration he remorselessly devoted half a century to the elaboration of his theory of "Oriental despotism." In Wittfogel's interpretation, modern communist societies fall into the same category of "hydraulic civilization" as the ancient Oriental societies, and all of them are located on the despotic side of a sharply defined frontier between freedom and coercion.

To accommodate his theory to historical facts (and vice versa) Wittfogel divided the "hydraulic world" into an array of subcategories: "loose" and "compact" hydraulic societies, "core" and "marginal" (and "submarginal"), three "complexity patterns," and various "density patterns" and "subtypes." Thus, despite traditional Russia's predominant dependence on rainfall agriculture, he was emboldened to include it in the "submargin of hydraulic society." This characterization of Russian agriculture and society as hydraulic (albeit marginally) lent itself to Wittfogel's interpretation of Russian and Soviet political history. He could now detect a menacing tendency in Russian social development towards "Oriental despotism," a tendency that he claimed was sharply exacerbated by the Bolshevik Revolution.

Although the concept of the Asiatic Mode of Production was defined and affirmed by Marx and Engels and never repudiated by them, it was rigidly proscribed by the communist movement for thirty years and has remained a tainted notion, suspiciously revisionist, among Marxists even after its partial rehabilitation in the 1960s. Its rejection was based more on political than on intellectual grounds, and the politicization of the concept has carried over into attempts to explain its renunciation. As a result, controversy and confusion abound here too.

Wittfogel, always ready to turn every argument into a vilification of the communism he had rejected, accounted for the repudiation of the Asiatic theory by accusing Lenin and his followers of attempting to conceal an "Asiatic restoration"—a state of despotism—that a communist revolution would bring about and which the concept of the Asiatic Mode of Production would reveal only too clearly (Wittfogel, 1981b: 390–400).

The British Marxist historian Eric Hobsbawm, reviewing the condemnation of the doctrine from a position 180 degrees removed from Wittfogel's, offered a more benign but no less political explanation: "The fear of encouraging 'Asiatic exceptionalism' and of discouraging a sufficiently firm opposition to (western) imperialist influence, was a strong, and perhaps the decisive, element in the abandonment of Marx's 'Asiatic mode' by the international communist movement after 1930" (Hobsbawm, 1964: 60n). Perry Anderson defended the faith with inflamed rhetoric that soon burned out of control. He justified the Marxists' rejection of the Asiatic concept for its "Confusionism," a "ubiquitous 'Asiatism'" that "defies all scientific principles of classification." In a long critique of the concept, he unloosed an expression of fury in his one reference to Wittfogel: "The most extreme form of this confusionism is, of course, not the work of a Marxist, but of a more or less Spencerian survival: K. Wittfogel, *Oriental Despotism*. . . . This vulgar charivari, devoid of any historical sense, jumbles together pell-mell Imperial Rome, Tsarist Russia, Hopi Arizona, Sung China, Chaggan East Africa, Mamluk Egypt, Inca Peru, Ottoman Turkey, and Sumerian Mesopotamia—not to speak of Byzantium or Babylonia, Persia or Hawaii" (Anderson, 1974: 486–87).[10] Scholars outside of the Marxist tradition have simply assumed, with less political passion, that the Asiatic Mode of Production was irrelevant to any analysis of European society because it did not occur in the European sequence—"the 'oriental epoch' was seen as outside the experience of Western civilization" (Service, 1978: 25).

Still another explanation of the rejection of the doctrine is that, in general, geographical and ecological interpretations of history and society are disowned by advocates of programs of political concern and action, because those interpretations may encourage what is known in some circles as "quietism," the attitude that whatever will happen will be determined by material conditions and cannot be influenced by organized political activity. Many Marxist scholars, deeply committed to social change, have heavily favored the political dimensions of Marxism over its full range of analytical possibilities. They have, accordingly, passed over the technological and ecological foundations of economic systems, and have begun their analyses on the sociopolitical level—the importance of property relations, invidious access to wealth and privilege, and monopolization of power by a dominant class—where collective political activity may be expected to have a salutary effect on the reform of society.

Despite the complexity of Wittfogel's thesis and its seemingly ad hoc categories, it has been recognized as a major achievement of interpretive scholarship. The competence, broad learning, and single-mindedness of his analytical and theoretical formulations, along with his close studies

of several hydraulic societies, have earned his work the respectful attention, and often the concurrence as well, of scholars in diverse fields.
Moreover, the high correlation between the development of pristine
civilizations and their dependence on hydraulicized agriculture in arid
zones has added measurably to the plausibility of his claims. One of his
critics conceded that "it cannot be a mere accident of history that so
many of the ancient major states should have started from a 'hydraulic
core.' Some common sociological principle must be at work" (Leach,
1959: 23).[11] The pungently anticommunist political thesis that Wittfogel wrote into his work, combined with his claim that communism has
indeed resulted in a restoration of Oriental despotism, has evidently
appealed to some readers on political grounds, and even fired the hope
that his work would produce desirable effects in the world of practical
politics.[12] Not least, the virtuoso performance of mind and body that
Wittfogel dedicated to his compulsively monographical studies from the
1920s until his death in 1988 resulted in a considerable extension, elaboration, and elucidation of his views.

With regard to the role of science in the ancient hydraulic civilizations,
Wittfogel essentially repeated the assertions Marx and Engels had made
about the importance of astronomy as an instrument of time reckoning
in societies whose basic economies rested on hydraulic agriculture:

Only when the beginning and ending of the rains, the rise and fall of the rivers,
becomes vitally important, does the need for a relatively precise calendar arise as
an immanent problem of the production-process. In general, where a leading
group concentrates in its hands the control of the waters, we find this group and
the state which stands behind it as well, in direct or indirect control of astronomy.
Thus two new types of productive forces are found [i.e., irrigation engineering and
astronomy] which give to the state functions which in other agrarian societies it
did not have to fulfill. (Wittfogel, 1981a: 147)

By generalizing these principles of social analysis it is possible to construct a model of scientific development in which geographical and ecological variables come into play. In many parts of the ancient world physical conditions did not warrant the development of food-producing
societies at all and, accordingly, in those regions of Paleolithic food collecting no urban development and no scientific cultures came into being.
Where, conversely, food production arose, it sometimes took the form,
depending on the landscape, of pastoral nomadism rather than farming.
And pastoral nomadism, like food collecting, produces no urban or scientific cultures. Only where intensified agriculture developed was the cultural scene markedly different. Early agricultural civilizations were generally situated in arid and semi-arid zones of the Old and New worlds;
variously designated as the Asiatic Mode of Production, Oriental despo-

tisms, or hydraulic civilizations, they fostered scientific cultures in which science was bureaucratically organized and was dedicated to the interests of the state.

But the ancient world also saw the rise of an agricultural civilization under divergent conditions. In contrast to hydraulic societies, Wittfogel cited those, such as classical Greece, where agriculture depended mainly on rainfall and where, consequently, no corvée was required to construct and maintain a hydraulic infrastructure, obedience to authority was not a primary virtue, and political life could be more fragmented, centrifugal, and dispersed as many social forces came into being to offset the power of the state. In classical Greece a scientific culture developed that in its social organization contrasted sharply with those in prior and contemporary hydraulic states. The individual savant replaced the state-employed scribe, and natural philosophy—the philosophy of nature—in the service of the individual's curiosity and intellectual passion replaced astronomy and astrology in the service of a centralized monarchy.

On the grounds of this bipolar historical sociology, in which bureaucratically centralized societies are contrasted with societies where individualistic efforts had greater scope, it should be possible to construct a binary model of the development of ancient science. The Oriental civilizations that Marx designated as the Asiatic Mode of Production would form one pole of the model, and Hellenic Greece the other. Later societies, whose traditions derived jointly from these two historical roots, would produce a variety of patterns in which characteristics of both traditions would be combined. On the one hand there would be individual savants dedicated to abstract theory; and on the other would be institutions supported and controlled by a central authority that deflected research and higher learning towards the interests of the state and the public.

SCIENCE AND ITS HABITATS

The history of science shows that the scientific enterprise alternately flourishes and subsides. It appears and endures, sometimes for centuries; it declines and sometimes disappears; it moves from place to place.

The pulse of science has been closely recorded. Traces of astronomical interest were left by Stone Age people. The great Bronze Age kingdoms that took root in Mesopotamia six thousand years ago patronized science and bureaucratized its pursuit, as did similar kingdoms in ancient Egypt, India, Ceylon, Southeast Asia, China, Mexico, and Peru. In the form of natural philosophy science arose in classical Greece, where it flourished for three hundred years before its center shifted, in step with Greek imperialism, to Alexandria in Egypt. In Islamic societies science fanned out east and west to Central Asia and to Spain. Modern science took hold in cisalpine Europe and then drifted first northward toward the coasts of the Atlantic, then into Central and Eastern Europe, and eventually westward across the ocean to North America. In our time it made another westward crossing, this time in the wake of waves of migrating political refugees. And as modern science established its reputation for utility, science and its institutions spread over the whole world in every direction from its European and North American bases.

The shifting patterns of scientific change are recorded in both time and space. Along with its history, assiduously studied, science also has a geography, still encrypted in unexamined assumptions and obscured by persistent disregard. Under the influence of a historiography of science that is mainly concerned with the internal logic of ideas, it is often assumed that geographical movements and distributions neither result from nor produce significant scientific change. Science may reveal fine shades of national or regional fashion, but unlike art and literature, which are stamped with distinct cultural styles, it is held to be a unitary entity, essentially alike wherever it appears. It may divide itself by content into the several sciences, and temporally into periods. It may lead or lag in specific cultural settings, but it always follows the same path, and it has finally all flowed together, forming a single cultural tradition that sprouted in the classical world of the eastern Mediterranean and has yielded essentially the same fruit, whatever the soil in which it has subsequently germinated. It may be viewed against a cultural or socioeconomic backdrop, depending on the leanings of the historian. But it is generally assumed not to change through the effects of its material settings, geographical or ecological.

We are thus left with the impression of a universal scientific tradition, altered over time, through a progressive development that may be punctuated by revolutionary transformations, but only minimally over space and material habitat. In regions where it has not kept pace with the advancing front of research and discovery, it is held to be merely retarded, for whatever reason, but following in the same track of the science whose development we know. Islamic science in particular has been poorly served by this linear historiography, which regards it as merely a conduit through which Greek science reached the Latin West.

Contemporary science has no doubt strengthened this impression of a single scientific tradition. Science has indeed become a unitary enterprise. Its principles are shared (or disputed) without regard to country, continent, or material conditions. When, conversely, a national, cultural, ethnic, or class distinction is claimed for contemporary science it is regarded with suspicion and usually repudiated with scorn. The institutions of science have everywhere been modeled on those of Northern

Europe; government support, heavily biased towards applied rather than pure research, follows a common pattern. All of these traits of science have been clarified by historical and sociological research. But in the many studies in which the characteristics of science have been plotted, only rarely has geography, treated as more than the name of a place or the site of a culture, formed one of the coordinate axes, despite the fact that, taking the world as the stage, scientific cultures have in the past differed sharply across geographical boundaries.

The disproportionate predominance of the history of science over its geography has resulted, in part, from a parochial interest in the development of science in a single geographical habitat, Northern Europe. Over the past few decades, comparative studies across diverse regions over long periods of time have begun to correct the imbalance between history and geography. But, ironically, the neglect of geography can easily be worsened by studies of non-Western science insofar as those studies may be motivated by an uncritical democratic impulse that is wary of highlighting uneven intellectual attainments among different populations in different habitats. To those who hold in high regard both egalitarian principles and scientific achievement (a sizeable cohort in educated circles), the opinion that some regions and habitats fostered little or no scientific development may appear to be intellectually and socially offensive and may, therefore, receive little attention. Art is generally recognized to have reached high levels of distinction in a wide range of societies, including some that were primitive and ancient. Technology, on the testimony of museums filled with stone tools, is the product of all incarnations of the genus *Homo*, and may even reach back to its australopithecine ancestors. But, for the democratic mentality, science, if not handled with care, may leave the unpleasant elitist impression that only Europeans excelled in its development.

Joseph Needham, whose in-progress study of Chinese science and technology has redrawn the profile of the history of science, has strenuously attempted to avoid that conclusion and has consistently interpreted the scientific enterprise in terms of both geographical diversity and Marxist internationalism. While conceding that "*modern* science was born in Europe and only in Europe," he declares that science is "the salutary enlightenment of all men without distinction of race, colour, faith or homeland, wherein all can qualify and all participate" (Needham, 1963: 149). Objecting to the "unjust" disregard of non-Western traditions, he made it clear that he meant not only that it was intellectually unjustified but also that it was uncivil, and morally and politically improper: "Unjust here means untrue and unfriendly, two cardinal sins which mankind cannot commit with impunity" (Needham, 1973: 1). As a sinologist Needham could confidently confront unsavory Eurocentric biases, for his chosen subject provides an example of a great non-Western scientific culture. But neither ecumenical admonitions nor the triumphs of Chinese science can resolve the inconsistency between a belief in the universality of science and the inevitable disclosure that, understood as the systematic and theoretical study of nature and numbers, science did not occur at all in many geographical habitats and that its development was stunted and enfeebled in others.

A study of scientific change that overlooks those vacant stretches in the history and geography of science will misconstrue diverse (and divergent) scientific traditions. These egalitarian concerns, however well meaning, focus attention on inappropriate questions. They emphasize similarities across geographical frontiers and minimize the differences that those frontiers define. They produce sympathetic accounts of each people's contribution to science, and they encourage attempts to explain away apparent failure. And they finally undermine a geography of science whose primary objective is not the assignment or denial of credit for accomplishment but an attempt to reach an understanding of correlations between habitat and culture. Stone Age food collectors, Neolithic farmers, and nomadic pastoralists of the Eurasian steppe, whatever naturalistic interests in heaven and earth they may have had, never created traditions of scientific research and never founded scientific institutions; while ancient Egypt, in proximate contact with societies like these but with an ecological base that led to intensive river-valley irrigation agriculture, gave rise to a highly developed urban civilization in which astronomy, mathematics, and oracular studies akin to science received considerable social encouragement and left records of substantial interest in research.

In recognizing these differences the geography of science is neither issuing a reproach nor granting approval; it is rather searching for cultural patterns that may depend on material conditions. The most meaningful implications of these recurrences and singularities are not to be found in any conjectures about intrinsic similarities or differences in the capabilities of various ethnic groups. It may be assumed that all human varieties are potentially capable of developing a scientific tradition. Despite this convergent potentiality, however, science has actually developed, or failed to develop, in dissimilar ways in diverse cultural and physical settings over its five-thousand-year evolution. The divergent settings, no less than the convergent potentials, promise to contribute to a further understanding of the structure and growth of science. It is the recurrences and singularities in the variety of material habitats in which science has been recorded or in which it apparently has not occurred that are the theoretical components of the geography of science.

The pattern of relationships between geography and culture is manifold. If science simply moves from one space to another without changing its environmental conditions it should change comparatively little. The mother-science of Europe was established in the environmentally similar eastern territories of North America as colonial science, and its development in the two locations remained nearly parallel with regard to both their cognitive content and their social organization. But the geography of science predicts that if science moves to a divergent physical habitat it will change more radically, as when science moved from the ancient Bronze Age kingdoms in Egypt and Mesopotamia to Ionia and peninsular Greece.

Although these correlations suggest that the material conditions of geography and ecology play a role in the development of science, those conditions have been resolutely disregarded in favor of historical, philosophical, and sociological perspectives. It may be that, when deployed in the process of social explanation, history is favored over geography because the former appears to allow more scope for volition and responsibility. Geography in social analysis raises the specter of determinism—geography is destiny. But weighing the effects of geographical factors on scientific cultures need not be inflated to an *ism*. It could be instructive, at least as a first approximation, simply to observe those effects, to correlate them, to study any patterns that may appear, and to appraise the results in the context of other efforts along these lines.

FRONTIERS IN SCIENCE

A preliminary sketch of the ecological frontiers that separate scientific cultures may be drawn by considering astronomy, the first science, in its prehistoric and early historic manifestations. Stone Age hunter-gatherers, Neolithic gardeners, and Bronze Age farmers, living under divergent material conditions, developed divergent traditions of astronomical interest. Thousands of lunar observations recorded by marks engraved on reindeer and mammoth bones and indicating a rudimentary interest in and knowledge of lunar cycles have been found dating back tens of thousands of years: ". . . evidence of lunar observation in notational sequences and markings dating from the Upper Paleolithic period; these extend backward in an unbroken line from the Mesolithic Azilian to the Magdalenian and Aurignacian cultures, a span before history of some 30,000 to 35,000 years" (Marshack, 1964: 743).

In contrast, among Neolithic food producers and fishermen living along the coasts and rivers of Europe, interest shifted from lunar to solar calendars, which provide greater accuracy in tracking the seasons. The instruments became architectural structures, instead of hand-held markers, culminating in elaborate and monumental calendrical devices like

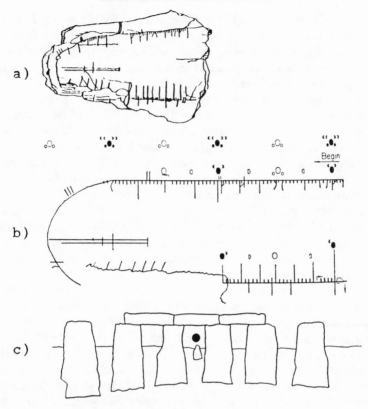

Fig. 1. Paleolithic and Neolithic astronomical devices
a) Engraved mammoth tusk (Upper Paleolithic) that has been interpreted as a record of
   lunar cycles. *Source:* Ščerbakiwskyj, 1926: 115.
b) Model of the lunar month aligned with the engraved markings. *Source:* Marshack,
   1964: 744. Copyright 1964 by the American Association for the Advancement of
   Science. Used by permission of AAAS and Alexander Marshack.
c) Neolithic observatory at Stonehenge. The sun rises directly over the heel stone on
   the morning of the summer solstice.

Stonehenge—stationary observatories more appropriate for sedentary
populations and requiring a more elaborate social organization, including
the patronage of the craftsmen responsible for constructing them. The
Neolithic communities that sponsored those works were presumably
even more interested than food collectors in discovering and recording
astronomical regularities that could reliably serve as a seasonal calendar
and, perhaps, as a navigational guide. In sedentary Neolithic societies
there was no longer any necessity that calendrical records be small and
portable; on the contrary, it was appropriate to make a calendrical device as
monumental as possible and to use it as a ceremonial center as well as an

observatory. A consortium of Neolithic villages could have provided the food surpluses to sponsor the construction and maintenance of such a rudimentary institution. But, neither lunar records inscribed on bits of bone or ivory nor observatories like Stonehenge reflect the formation of a developed scientific culture complete with cadres of scribes and scholars.

Later, a third development, beyond Paleolithic moon-gazing and megalithic observatories, occurred in the alluvial valleys and flood plains of the Middle East. There agriculture, intensified by natural and artificial irrigation, sustained urban civilizations, and written calendars were devised and staffs of astronomical specialists employed, not only to announce the march of the seasons, but also to divine the future in the service of court and temple patrons.

This progression of paleoastronomy over a span of thirty-five thousand years cannot be understood simply in terms of intellectual inspiration (or retardation), or solely by the fortuitous activities of outstanding individuals. Cognitive originality may account for the realization that solar solstices and equinoxes can be used to mark the seasons; but it cannot account, over thousands of years, for the making of calendrical devices, for passing them on from generation to Neolithic generation, or for employing them as ceremonial centers. Nor can it account for the royal patronage of astronomer-scribes in the first kingdoms. The act of discovery alone cannot explain the divergent patterns of recording and applying astronomical knowledge in Paleolithic, Neolithic, and Bronze Age societies. Sociological processes in diverse and changing ecological environments inevitably played a role in conjunction with the conceptual insights that were achieved.

The lesser development of astronomical knowledge in food-collecting societies and the failure of those societies to institutionalize astronomical observation cannot, on present evidence, be attributed to any mental incapacity. Over the past 200,000 years those societies have been the work of human beings whose varieties, if they differ at all in mental capacity, appear to differ only minimally. All ethnic taxa of man have at some time formed food-collecting communities, but none of these communities, regardless of the cognitive power of its members, ever developed an astronomical tradition comparable to those of agricultural societies. These facts alone indicate that the divergent traditions of gaining, recording, and preserving astronomical knowledge in primitive and early historical societies must be accounted for primarily in terms of social organization rather than mental power or biological heritage. In the prehistorian's elongated perspective the structure of society is, in substantial measure, determined by ecological conditions. Indeed, if recent research is confirmed, the fateful transition from hunting-and-gathering to food-producing economies was the direct result of neither an agricultural invention nor any transcendental revision of man's relationship with nature; it was sparked instead by misalignment among ecological variables, primarily increasing population density and diminishing availability of collectible plants and large-bodied animals.[13]

Few prehistorians have delved into the question of science in Paleolithic and Neolithic societies; and the most eminent among them, Gordon Childe, reached conclusions that are inconsistent with the formulations of most historians of science. Viewing the issues from a soft-core Marxist position, Childe conflated science and technology, thinking of both as applied science, and was thereby led to believe that science, which he often designated as "practical" or "contemporary" knowledge, was a common feature of the Paleolithic scene:

Like any modern machine or construction . . . ancient relics and monuments are applications of contemporary knowledge or science existing when they were fashioned. In a liner [i.e., an ocean liner], results of geology (oil, metal-ores), botany (timbers), chemistry (alloys, oil-refining), and physics (electrical equipment, engines, etc.) are combined, applied, and crystallized. That is equally true of the dugout canoe fashioned by Stone Age man from a single treetrunk. (Childe, 1951: 13)

Similarly, with regard to the Neolithic, he asserted that, "a true loom goes back to neolithic times. . . . Its inventors . . . made an essential contribution to the capital stock of human knowledge, an application of science" (Childe, 1951: 80). These were not merely rhetorical solecisms. Childe firmly believed that prehistory witnessed a pristine identity of science and technology: "We have shown in detail that [science] originated in, and was at first identical with, the practical crafts" (Childe, 1951: 179). By casually equating science and the crafts Childe obscured and confused the correlations among science, technology, and society. The bow and arrow undeniably embodies accurate and useful knowledge gained through direct experience of natural processes. The inventor of the bow may have learned the hard way that bent twigs snap back viciously when released; but Galileo, without ever bending a twig or a cantilever, discovered by means of deductions from mathematical statics the surprising (and, at the time, useless) principle that the strength of a bent beam is proportional to the square of its depth. Knowledge directly gained from experience and directly applied to technique is essentially unlike knowledge derived through abstract thought with, frequently, no foreseeable application. The auto mechanic possesses a wealth of knowledge, but what he knows is different from what is known by a thermodynamicist (who may be incompetent to repair an engine). These divergent forms of knowledge are both cognitively distinct and connected to society and its material base in different ways. The auto mechanic and the thermodynamicist, egalitarian ideals notwithstanding, are separated by sociological barriers as well as by intellectual traditions and habits. If no such barriers existed in Paleolithic bands or Neolithic villages it was only because there were no individuals comparable to research scientists. Surely no Paleolithic craftsman can be reasonably characterized as a scientist; and science without scientists is merely a conundrum, not a principle of historical or anthropological analysis.

In retrospect it is always too easy to be ensnared by the assumption that the scientific principles and procedures that are now employed to understand a given technique somehow contributed to the discovery of the technique. But that contribution cannot be assumed. It must be demonstrated by showing that the principles or procedures had been formulated prior to, or perhaps in the course of, producing the technical innovation, and that they had actually been applied in its development. Moreover, if the use of "timbers" (in Childe's account) necessarily confirms the application of botany, or the digging of "metal-ores" the science of geology, then science, in its cognitive sense, would dissolve into a vague, often utterly unconscious process that would inevitably accompany every purposeful human action. The bow and arrow would be revealing of theories of elasticity and ballistic motion, and even the chipping of stone tools would be equated with geology and the strength of materials.

Benjamin Farrington, a British Marxist historian who was influenced by Childe and who also referred to "the science implicit in techniques," recoiled from these implications and refused to muddle science so flagrantly. He conceded that while the "spear thrower, the bow and arrow, the bow-drill, are all so many landmarks in [Stone Age man's] progress in mechanics . . . his appreciation of the principles involved is at first practical, sensuous, merged in the operations, untheoretical" (Farrington, 1980: 16, 21). Childe, too, seemed at times to recognize the fallacy of uncritically equating science and technology. He softened his claim that they are equivalent by referring to "sciences of a new kind," which appeared in the urban-centered, civilized societies of the Middle East; and he even allowed that "the distinctive achievements of civilizations that differentiate them from barbarism are the invention of writing and the elaboration of exact sciences" (Childe, 1942: 132–33).

Although Childe's attempt to equate science with the crafts has received no sound confirmation, it continues to inspire conviction. In a close and sympathetic review of Childe's work, Bruce Trigger stated: "He not only demonstrated that archaeological data might constitute a major element in an expanded and enriched history of science, but also carried out more substantive research in this area than in any other. While prehistoric archaeology in general has not developed closer links with the history of science since Childe's time, his proposal remains worthy of further consideration, especially as a result of the development of industrial archaeology" (Trigger, 1980: 179). Trigger's hopes for industrial archaeology simply repeat Childe's mistake by replacing the unsubstantiated equivalence of science and technology with the unwarranted affiliation of science and industry. All of the examples of technical progress that Childe discussed were benchmarks in the development of technology, not science, at a time long before the two activities began to merge, and even before scientific cultures can be clearly identified. Rudimentary features that are of scientific interest can perhaps be seen in the Paleolithic recordings of lunar cycles and in the astronomy embodied in megalithic stone rings; but prehistory, the long era prior to the rise of urbanized, state-level civilizations in the ancient Middle East, was devoid of activity that can properly be classified as a developing scientific tradition. The merger of technology and theoretical science is not inherent in those activities. It is a historical process that has occurred under specific conditions, and its description is a problem for historical research.[14]

Perhaps it was in an attempt to raise the status of the crafts by equating them with science that Childe postulated scientific developments in social and geographical settings where they had not occurred. The coupled notions that technology is applied science and that science is inherently useful in an economic sense are common misconceptions that evidently have wide appeal. If science is placed among economically useful activities it need not be merely relegated to an elitist cultural superstructure; and if the crafts are science, then non-Western nations and nonlearned social groups can be credited with substantial intellectual accomplishments. Since technology is conceded to be common to all societies, the association of science with technology reinforces Needham's idea of a universal science "wherein all can qualify and all participate." But, however agreeable the equation of science and technology may be to a fraternal sensibility, it is analytically pernicious. It obscures the patterns of change that have characterized the growth of science; and, by seeing science everywhere, it has without reasoned arguments restricted the study of a geography of science that attempts to trace the origin and profile of scientific development in divergent material habitats.

The frontiers between proto-astronomy in Paleolithic and Neolithic communities and the more robust practice of research in economies based on intensified agriculture coincide with geographical boundaries between ecological zones. The river valleys and flood plains of the Middle East where Bronze Age kingdoms originated and where written scientific traditions were first systematically recorded are ecologically and culturally distinct from territories where food collecting or incipient farming remained predominant. For Childe and Farrington the cultural frontier between the craft knowledge of Paleolithic food collectors and Neolithic horticulturalists, on the one hand, and the developed calendrical astronomy and oracular astrology of civilized societies, on the other, is a frontier between "practical science" and "learned science." But once the confusion over science and technique is eliminated that frontier becomes the boundary that marks the beginning of science. Ludwig Edelstein emphasized the distinction that Childe and Farrington had obscured: "The question when science began may at first glance seem as meaningless as the question when poetry and art had their beginning. Yet provided that by science one does not understand just any kind of knowledge or technical skill . . . but a consistently rational explanation of phenomena, the question is meaningful and allows of an answer" (Edelstein, 1957: 113–14).

At various times and in several places the cultural frontier demarcating the origin of a scientific tradition coincided exactly with the geographical frontier separating the intensive, irrigated agriculture and urban civilization of settled farmers, generally in river valleys, from the "barbarian" communities—based on hunting and gathering, pastoral nomadism, small-scale rainfall agriculture, or a combination of these economies—that dwelled on their borders.[15] On the barbarian side of the frontier there was considerable knowledge of natural and artificial processes but little systematic study of nature, few traces of a scientific tradition, and no scientific institutions; on the other side, in the great civilizations of Egypt, Mesopotamia, India, China, Mesoamerica, and Peru, there were temples and courts in which clerks and scribes were employed to study calendrical astronomy, to delve into numerical and astrological divination and prognostication, to derive methods of calculation, and to study and practice medicine and surgery. If these activities had not yet acquired all of the earmarks of science, they were at least "akin to science (in a kind of ancestral relationship)" (Lounsbury, 1978: 804). Scientific interests were recorded—on papyrus, on clay tablets, on tombstones and walls—and generated continuous traditions of study; they were patronized by royal and sacred institutions and have come down to us as evidence of the dawn of science.

The bureaucratic encouragement of science in the first kingdoms and its absence or relatively slight development in Neolithic communities are consistent with economic possibilities and requirements in geographically diverse regions. Stonehenge and the other megalithic rings of ancient Britain and Ireland reflect a rising level of astronomical interest and knowledge consistent with a stage of Neolithic fishing, hunting, and gardening that could produce surpluses sufficient to sustain small contingents of craftsmen but too small to support an urbanized civilization.[16] Although the largest European megaliths may have embodied as much labor as some of the monumental architecture and hydraulic structures of Egypt and Mesopotamia, they did not entail the constant effort, year in and year out, required to maintain an irrigation system; they did not intensify the production of food the way hydraulic agriculture did; and they did not call for the allocation of water and the taxation and redistribution of food surpluses by a central authority. Compared to the high civilizations of the Middle East, prehistoric Europe, although the scene of settled agriculture (or at least horticulture), was a region of low population density, no large urban concentrations, small food surpluses, and no tendency toward bureaucratic centralization. Under those circumstances astronomical observation might reach an impressive level of development, as it evidently did at Stonehenge and similar sites, but there could be no transition to a new cultural pattern of government-sponsored legions of scholar-scribes.

Geographical frontiers of this type, between regions where science was systematically sponsored by bureaucratic patrons and regions where, contemporaneously, it was absent or at most thinly diffused and lacking an institutional base, have appeared repeatedly where settled farmers in river valleys and on flood plains supporting an urban civilization bordered Paleolithic food collectors, Neolithic gardeners, or steppe nomads.[17] The first of these frontiers was crossed in the ancient Middle East.

EAST IS EAST AND WEST IS WEST

The Mediterranean littoral is for the most part arid, the more so as one moves from west to east and from north to south (Braudel, 1975, vol. 1: 235; Semple, 1931: 90). Still farther east, across the Arabian peninsula, the Mesopotamian flood plain is held in the same parched regimen. Along the Nile in Egypt and the Euphrates and the Tigris in Mesopotamia annual rainfall amounts to less than twenty inches, and frequently less than ten inches. And yet, these two desiccated regions, where dryland farming scarcely is possible, occupy the better part of what is paradoxically known as the Fertile Crescent.[18] The fertility of this crescent-shaped territory and the resolution of the paradox depend on irrigation and flood control, which in turn have for millennia depended to a greater or lesser degree on the hydraulic engineering works required to construct and maintain catchments, canals, conduits, drainage channels, sluices, terraces, and embankments.[19]

In Egypt, irrigation ditches watered the lower valley and distributed fertilizing silt in the form of pulverized volcanic rock brought by the Blue Nile from Ethiopia and as organic material brought by the White Nile from central Africa. In Mesopotamia, the nutrient largesse came down from the highlands of what are now northern Iraq and Syria and eastern Turkey, and resulted in such quantities of silt being deposited that the rivers were often driven from their beds.[20] In both cases, natural and artificial irrigation transformed desert into lush, fructiferous soil sustaining the most intensive agriculture and the densest populations in the early historic era—the fabulous Oriental civilizations of the Middle East with their monumental structures, their splendid works of art, their famous cities, their learning, and their despotism. The art and architecture took form in statuary, wall paintings, temples, tombs, and palaces;[21] the despotism was centered in a pharaoh or king; and the learning embraced knowledge that ranged across the fields of calendrical reckoning, algebra and geometry, astronomy and astrology, chemistry, medicine, and dentistry. The substance of the learning has been closely studied and has revealed some of the singular characteristics that distinguish it from both the astronomical observations of precivilized populations and the science that was later created by the Greek-speaking populations of the eastern Mediterranean.

Most of the discoveries recorded were evidently made by scribes or physicians patronized by political or religious officials and employed as civil servants in temples, scribal schools, or palaces.[22] In Mesopotamia, scribal schools devoted to the interests of governance were already founded in the earliest civilizations and functioned not only as teaching institutions but also as repositories of higher learning and to some extent as research centers:

The original goal of the Sumerian school was what we would term "professional"—that is, it was first established for the purpose of training the scribes required to satisfy the economic and administrative demands of the land, primarily those of the temple and palace. This continued to be the major aim of the Sumerian school throughout its existence. . . . Within its walls flourished the scholar-scientist, the man who studied whatever theological, botanical, zoological, mineralogical, geographical, mathematical, grammatical, and linguistic knowledge was current in his day, and who in some cases added to this knowledge. (S. Kramer, 1959: 2)

In Egypt, too, institutions of learning are referred to in the ancient texts. The "House of Life," which is first mentioned in the Old Kingdom, functioned principally to maintain political order in the interests of the state. In addition to attending to ritual customs, it was also "a scriptorium for the production of many kinds of works [including scientific treatises]," and it sponsored studies of magic, medicine, astronomy, and mathematics. Besides "houses of life" there was a record hall, a house of books, a "hall of writing," and temple libraries (Clagett, 1989: 25, 32–35).

In both Mesopotamia and Egypt the scribal profession was held in high esteem and maintained as a preserve of the upper class. In Sumer the ranks of the profession were filled by the sons (no daughters) of the "wealthier citizens of urban communities" (S. Kramer, 1959: 3); and in Egypt there was a strong correlation between the scribal profession and the achievement of high status as a government official.[23] The importance of writing in a highly centralized state that was responsible for the management of the basic agricultural economy undoubtedly accounts for the importance that was attached to scribal learning.[24]

Judging by the numerous results that have been attested for both Egypt and Mesopotamia over many hundreds of years, there were presumably many individuals thus employed. Yet, surprisingly, for neither Egypt nor Mesopotamia has it been possible to construct a scientific biography of any individual savant who contributed to science. Over those long stretches of time, the night sky was tracked, calendars were devised and refined, the volumes of prisms and of truncated pyramids calculated, chemical processes studied, roots of quadratic and cubic equations extracted, Pythagorean triplets listed, and irrational and transcendental numbers approximated, yet the court and temple scholars who made those discoveries remain invisible to the most searching biographical inquiry. The earliest entry in the authoritative *Dictionary of Scientific Biography* (*DSB*), with its six thousand articles, is the Greek thinker Thales, who died in the sixth century B.C. (Gillispie, 1970–1980).[25]

Nor can it be the mists of time alone that have obliterated all trace of the scribes who did the research and made the discoveries in the fields of science, for, among architects and practicing physicians, individuals have been identified and have even left their autobiographies.[26] The anonymity of the Oriental savants is thus no artifact of the failure of historical investigation. It may be that the personalities were condemned to obscurity because research and the discovery of new knowledge were shrouded in deliberate secrecy for reasons of state.[27] But whatever the reasons, the biographies of those ancient sages remain unwritten. The explanation of the anonymity of multitudes of learned experts is itself a secret, concealed in the bureaucratic social organization of the centralized civilizations that developed in the river valleys of the ancient East.

The organization of science as a bureaucratic civil service and the anonymity of its practitioners were two distinctive features of the ancient Oriental scientific tradition. In addition, the state-sponsored research was focused on the utilitarian interests of its bureaucratic patrons. Calendars tracked the agricultural seasons and marked ceremonial events, astrology designated "lucky days" for planting crops and guided the monarch in affairs of state, geometry was applied in the design and construction of pyramids and canals, and even seemingly abstract mathe-

matical knowledge had distinctly practical objectives.[28] Otto Neugebauer, the eminent authority on the mathematics of the ancient Near
East, observed that

many of our mathematical tables are combined with tables of weights and measures which were needed in daily economic life. There can be little doubt that the
tables for multiplication and division were developed simultaneously with the
economic texts. Thus we find explicitly confirmed what could have been concluded indirectly from our general knowledge of early Mesopotamian civilization. (Neugebauer, 1957: 31)

Even elegant theoretical research had a utilitarian slant in the direction
of state interests. Egyptian scribes somehow hit upon a procedure for
calculating the volume of a truncated square-based pyramid, a result that
might have been used in their monumental engineering projects.[29] The
famous pyramids were of course truncated until the final phase of their
construction. Calculations of their increasing volumes would, therefore,
have been useful in determining the amount of work done, and thereby
the contractors' fees, as the work progressed. It would have been a great
convenience to calculate the changing volume simply by measuring a leg
of the lower base (which was known from the start), a leg of the upper
(truncated) base, and the height to the level of truncation (which was
readily determined by counting the steps).[30]

Similarly, for Egyptian astronomy it has been affirmed that "to a very
high degree [it] was the severely practical servant of Egyptian time-
reckoning. This was true over the long span of Egyptian written history"
(R. Parker, 1978: 706). Again and again, authorities on the sciences of
ancient Egypt and Mesopotamia have noted their "applied" character,
narrowly related to technical tasks, and their corresponding indifference
to the abstract, speculative thought that, following the Greeks, we have
come to expect as an essential component of the study of nature and
numbers.

Only occasionally do recorded research results seem to reflect the purely abstract interests of a scribe. Just as the applied sciences today sometimes poach on the territory of pure research, it was probably inevitable
that, in the course of systematic studies of arithmetic and geometry,
relationships would occasionally be discovered that had no immediate
application. Some ancient Babylonian scribe, perhaps indulging an irre-
pressible urge to play with numbers, approximated the square root of 2 to
the equivalent of six decimal places, beyond any conceivable need in
engineering or reckoning. Even this step towards pure mathematics was
taken in the context of broad programs of study directed at practical ends.
Tables of exponential functions that would appear to be as abstract as an
excessively accurate approximation of the square root of 2 were in fact

used to calculate compound interest. Quadratic equations were solved in connection with the deployment of labor. Linear equations were solved to determine shares of inheritance and the division of fields. Lists of coefficients for building materials may have been used for the quick calculation of carrying loads. Coefficients for precious metals and for economic goods presumably had equally practical applications. And calculation of volumes reflected no abstract interest in geometry but was applied in the construction of the hydraulic infrastructure (Neugebauer, 1957: 34–35, 42–46): "Whole sections of problem texts are concerned with the digging of canals, with dams and similar works, revealing to us exact or approximate formulae for the corresponding volumes. But we have no examples which deal with these objects from a purely geometrical point of view" (Neugebauer, 1957: 46). Under the demanding patronage of the ancient autocrats, higher learning was compulsively dedicated to the solution of practical problems.

Along with this utilitarian orientation, and possibly integral to it, was a proclivity to treat knowledge extensively, by drawing up tables of numbers and making lists of objects. This manner of coping with and recording knowledge—what has been called the "science of lists" (Taton, 1963: 75–78)—may have been favored generally in societies that had not yet discovered formal logic and analytical thought. The laborious drudgery that went into it, intellectually unrewarding to the individuals who compiled the data, may have been possible only where battalions of scribes were patronized by the state as civil servants. But, whatever its meaning and its social significance, listmaking gave Oriental learning a distinctive identity in both Mesopotamia and Egypt where scribes compiled an indiscriminate variety of lists—of things, words, numbers, gods, cities, occupations, and scribes. Orientalists have puzzled over this caprice of scribal mentality and have tried to explain its origins. Robert McC. Adams observed:

Among the earliest pictographic tablets from Uruk are not merely crude records of economic and administrative transactions but formalized lists of gods, professions, geographic names, and classes of objects arranged in conceptual categories. . . . The list-keeping, classificatory aspect of cognition that permanently stamped scribal learning for as long as there was a cuneiform script thus had made its full-blown appearance here very soon after the script's origins, perhaps reflecting long familiarity with simpler counting and mnemonic devices. (Robert Adams, 1981: 80)

But despite its kinship with formal, if stultified, learning, listmaking has been distinguished from science proper:

It cannot and should not be claimed, of course, that the word lists containing, for example, the names of plants, animals, or stones constitute the beginnings of botany, zoology, or mineralogy in Mesopotamia. They are not a scientific (not even a prescientific) achievement; rather, they result from a peculiar interaction of a genuine interest in philology (or, at any rate, lexicography) and a traditional Near Eastern concern for giving names to all things surrounding the scribe, thus linking nature to man. (Oppenheim, 1978: 636)

A. H. Gardiner, who devoted many years to the study of ancient Egyptian lists (onomastica), drew attention to the scribes' rudimentary attempts at classification, attempts that were occasionally indifferent to jarring inconsistencies and incommensurate categories. Commenting on a list of "entities in the physical world," Gardiner observed:

Even if we were to accord to the entire work as many as 2,000 items—a number probably much in excess of the truth—such a figure would be fantastically small for a catalogue of the universe, the more so when it is noted that the honour of a separate mention is done to particular kinds of pastries or cakes . . . which thus receive as much individual attention as the great city of Memphis . . . or as heaven itself. . . . Out of such grotesque beginnings have our encyclopedias arisen! (Gardiner, 1947: 35)

The diversity of these modern accounts of an ancient oddity testifies to both the extensive ancient practice of compiling lists and to the uncertain state of expert knowledge today about why the scribes practiced that peculiar art. It may be that since making and keeping lists of commodities for economic reasons was one of the primary functions of court and temple scribes, they reflexively applied those skills and habits to a great variety of objects, not always with clear and specific utilitarian purposes.[31] Whatever the ultimate explanation of this general tendency to cope with knowledge by tabulating data, it provides one more marker along the frontier between the higher learning of the Oriental civilizations and the analytical science of the Greeks.

The astronomy of the ancient Oriental kingdoms is legendary, and its utilitarian bias is unmistakable. In Babylonia the earliest astronomical texts (c. 1830–1531 B.C.) were concerned with astrological predictions of agricultural outcomes (van der Waerden, 1978: 672). And in Egypt, where science was in general less robustly developed than in Mesopotamia, calendrical reckoning was brought to a high level of refinement.[32] Neugebauer asserted that the Egyptian calendar, which "originated on purely practical grounds," is "the only intelligent calendar that ever existed in human history." It abandoned any attempt to correlate lunar cycles with the solar year, and it calculated the year as twelve equal months of thirty days each and five additional "leap-days" (Neugebauer, 1957: 81). Consistent with the requirements of an agricultural society basically dependent on the timely rising of the river that fed its life-sustaining irrigation system, the Egyptian calendar originated as a service to agriculture rather than out of any theoretical interest in astronomy. "Its basically non-astronomical character is underlined by the fact that the year is divided into three seasons of four months each, of purely agricultural significance. The only apparent astronomical concept is the heliacal rising of Sirius which, however, obtained its importance only by its closeness to the rising of the Nile, the main event in the life of Egypt" (Neugebauer, 1957: 82).

In contrast with the utilitarian bent of calendrical astronomy in the East, Greek astronomy was to become less subservient to time reckoning and would acquire a new position as the ornament of a secular natural philosophy dedicated to abstract learning and dissociated from practical applications. In the Bronze Age kingdoms of the Middle East, however, scribal learning was not focused on any abstract study of nature. It was essentially practical. It was thoroughly entangled with religion and magic and was sharply skewed towards prognostication and the discovery of portents, generally in the service of the state: "The main motivation for Mesopotamian man's interest in keeping the manifestations of animal and plant life, the movements of the heavenly bodies, and other phenomena under close and constant surveillance was his hope to obtain from them timely warnings of impending misfortune or disaster" (Oppenheim, 1978: 641). Similarly, cosmological ideas and the attributes of deities were related to practical interests, often in connection with the natural and ecological environments. In Egypt, the "various cosmological systems also share, no doubt, their primitive origins in the two pervading natural features . . . : the overwhelming importance of the Nile and its annual flooding and the ever-present sun as a continuous source of light and heat" (Clagett, 1989: 265–66); and in Mesopotamia, "the regularity of agricultural and sacral activities is the sole concern of the gods" (Oppenheim, 1978: 639).

All of these traits of ancient Oriental science were to be modified in classical Greece. Just as geopolitical states may face multiple frontiers, so too may a scientific culture. The river-valley civilizations of the ancient Orient were bordered not only by barbarians bereft of scientific traditions and institutions but also, late in their histories, by a variant civilization to the west. The "education of Hellas" was based, like the cultures of the East, on settled agriculture; but it arose on lands less arid than the alluvial valleys of Egypt and Mesopotamia, less dependent upon irrigation in its agricultural economy, less governed by centralized monarchies and bureaucracies, more dispersed and centrifugal. And it produced a scientific culture—"a dazzling light, a fearful storm"—so prodigiously remarkable that it has sometimes seemed to have had no antecedents and even now continues to challenge historical interpretation.

The splendor of the Greek achievement in science has divided historians into "lumpers" and "splitters."[33] Some have claimed that Greek science would be unaccountable unless it were derived from the neighboring civilizations of Egypt and Mesopotamia; others insist that its singular intellectual quality bespeaks a discontinuity between unlike categories.

The case for the splitters was stated by John Burnet:

The very fact that for two or three generations Greek science remained in some respects at a very primitive stage affords the strongest presumption that what came to Hellas from Egypt and Babylon was not really rational science. . . . [If] we mean by science what Copernicus and Galileo and Kepler, and Leibniz and Newton meant, there is not the slightest trace of that in Egypt or even in Babylon, while the very earliest Greek ventures are unmistakably its forerunners (Burnet, 1914: 4–5).

While Burnet's argument was only descriptive, not explanatory, some splitters have gone further and have tried to explain the postulated discontinuity between Oriental and Greek scientific traditions by attributing it to "instinct" or "race." Thomas Heath, a historian of mathematics, saw in the Greeks of antiquity a "genius for mathematics [that] was simply one aspect of their genius for philosophy. . . . The Greeks, beyond any other people of antiquity, possessed the love of knowledge for its own sake; with them it amounted to an instinct and a passion." He concluded that "the Greeks were a race of *thinkers*" (Heath, 1921: 3, 6). More recently, the apparent bimodality of Greek and Oriental scientific thought led Derek de Solla Price to speculate, with imagination and some abandon, about corresponding physical differences in the human brain:

Think, for example, of the Mayan, Hindu, and Babylonian art works with their clutter of content-laden symbolism designed to be read sequentially and analytically, and compare it with the clean visual and intuitive lines of the Parthenon! Strangely enough, it has now emerged from . . . psychological researches . . . that the difference in styles corresponds very closely with that of the activity of the left and right hemispheres of the human brain. The left hemisphere, controlling the right half of the body, seems to be "Babylonian," the right hemisphere and left half of the body "Greek." (D. Price, 1975: 22–23)

Lumpers (and probably most splitters as well) have understandably drawn back from such declarations. George Sarton, one of the founders of the study of the history of science in the United States, reproached "uncritical friends of Hellas," insisting that "early science was genuine and admirable, some of it was on a higher level than early Greek science. It is unfair to exaggerate the irrational aspects of the early Oriental science and to compare them with the most rational aspects of Greek science" (Sarton, 1952: 128). And Farrington chided Thomas Heath for his baseless ethnic characterizations: "We have acquired a distaste for explaining mental characteristics on a racial basis and . . . in any case, the Greeks were not a race but a people of mixed descent" (Farrington, 1980: 13).[34] Generally, the lumpers have only contested the extremism of the splitters' arguments without questioning the validity of recognizing some distinction between the characteristics of Greek scientific thought and Oriental accomplishments in the fields of astronomy, mathematics, medicine, and occult learning.

This unresolved controversy has not been confined to the internal, cognitive dimension of science. Sociologists and social historians, looking beyond the rational structure of knowledge at ways in which socioeconomic, ideological, and institutional developments have jointly affected scientific change, have also confronted the puzzle of Greek science. Joseph Ben-David, reviewing the history of early science with an eye toward the social role of scientists, hedged on the choice between lumping and splitting. He placed "ancient Mesopotamia, Greece, and China" in a single category in terms of both their "impressive achievements" and the "special role of the natural philosophers in Greece [which] was . . . similar to those of other ancient civilizations" (Ben-David, 1971: 22, 35).[35] He allowed that the "only difference" between the role of Greek scientists and that of scientists in other ancient civilizations lay in "the better preservation of the personalities and the teachings of these thinkers"; and he attributed the preservation to "the better quality of the people and their thinking, to special conditions, or, most likely, to the interaction of both." These "better people" were attracted to science by the "more favorable" conditions that "apparently existed in the case of the Greeks." What these conditions and their consequences might have been are only hinted at: "They were a dispersed and politically decentralized nation" (Ben-David, 1971: 35).

These unsubstantiated ethnic differences, far-fetched neurological conjectures, and obscure allusions to "better people" and "special conditions" reflect the difficulty of accounting for the contrasts between the ancient Oriental and Hellenic scientific cultures. Although many acute descriptive studies have been produced that contribute to an understanding of these differences, there have been few, if any, systematic attempts to correlate the cultural differences in the organization of these ancient scientific traditions with the ecological zones and corresponding material civilizations in which they flourished.[36]

The transition from the ancient Oriental world to the world of classical Greece was quite unlike the transitions from one period to another that historians commonly study. Specifically, it was different from the transformation of feudalism into capitalism, which, under the influence of Marxism, has often served as a model in studies of social change. In that case, the development from one social system to the other occurred in the same physical habitat, Western Europe. But the Bronze Age civilizations and Greek culture developed in two sharply divergent zones—the intensively cultivated river valleys of the Near East and the fragmented plains, coastal districts, separated valleys, and islands of the Aegean. While in Western Europe the emergence of capitalism out of the "womb" of feudalism cannot be attributed to divergent material settings (although such geographical factors as demographic growth and habitat degradation come into play), in comparative studies of the ancient Orient and Hellenic Greece geographical differences provide a clear opportunity to test the correlations between scientific cultures and material life.

The contrasting characteristics of Oriental and Greek science correspond to distinct physical habitats. The move westward from the Near East, where Oriental science was cradled, to the region of the Aegean Sea leaves behind the river valleys of Egypt and Mesopotamia, with their compact civilizations huddled along colossal rivers and their networks of embankments and irrigation ditches. In place of streams and canals is a sea with hundreds of islands and sharply indented coasts; in place of aridity is sufficient rainfall to make irrigation an agricultural benefit rather than a strict necessity; in place of dense civilizations physically hemmed in by sharp ecological limits was a centrifugal civilization dispersed by seafaring, commerce, emigration, and colonization. And in place of Pharaoh, with his irrigation engineers and his labor battalions, were independent city-states. In the absence of any compelling vital economic and technological functions that it had to perform on a continuous basis, comparable to the maintenance and control of an irrigation network and the redistribution of its products, the Greek state was, in comparison with the centralized governments of the East, a slack authority deficient in imperiousness, opulence, and absolute control.

In the Aegean environment, the locus of science moved from palaces and temples to the private quarters of solitary scholars or, occasionally, to the private schools they founded.[37] There were no longer clerks and scribes employed as civil servants. Indeed, there was no employment at all. Lacking private wealth or a wealthy patron the scientist was reduced to the status of a penurious garret-intellectual. The predicament of Hellenic scientists has been summarized by Ludwig Edelstein:

Unrecognized by his contemporaries the scientist found himself outside the confines of the social order. There were no jobs for him; science was not a career. He had to have independent means with which to carry out his investigations or to travel for the sake of his geographical studies. Society had a place for the practicing physician, but not for the biologist. . . . As interest increased in knowledge as such, the scientist began to teach others. . . . Occasionally he found a protector who provided for his livelihood. On the whole, even in the classical centuries everything was left to the initiative and courage and determination of individuals who pursued their studies at their own expense and at their own risk. Athens forced her citizens to pay for producing the plays of Aeschylus, Sophocles and Euripedes. At the great festivals everywhere, poets and musicians competed for rewards and prizes; no similar opportunity existed for scientists. (Edelstein, 1957: 114)

Paradoxically, the Oriental societies, which valued their sages for the practical benefits expected from their research, condemned them to impenetrable anonymity, while in classical Greece, which spurned its scientists, they achieved everlasting renown. Bureaus staffed by nameless adepts on one side of the frontier, egocentric individualists vying for acclaim and immortality on the other.

The internal, cognitive features of Greek science displayed the same contrast with Oriental science as the external, sociological features. Learning in the river valleys of the East was embodied in lists; Greek science was analytical. Logic, with the capacity to generate intellectual systems from a handful of assumptions without any further need to investigate the natural world, produced knowledge intensively, in contrast with the Oriental habit of listmaking, which recorded knowledge extensively. And where Eastern science was "applied" and divinitory, its objectives mainly practical, Greek science embraced a large measure of abstract, speculative thought—it was the philosophy of nature "symbolized in Democritus' saying that to find an etiology is greater gain than to become king of Persia" (Edelstein, 1957: 114). At each point, the two scientific traditions can be correlated with the sociological differences between an individual thinker wondering about nature and operating independently of any institutional constraints and a civil servant working on an agenda written by court and temple overseers.

Additional contrasts between the Greek and Oriental traditions can be drawn. Scientific cultures similar to those of the ancient Near East have appeared repeatedly, and sometimes independently, in different regions of the world—wherever high civilizations rose on a base of hydraulically intensified agriculture. In contrast, Greek science originated once—in the eastern Mediterranean. Oriental science favored arithmetic and arithmetical astronomy while Greek science displayed an underdeveloped arithmetic along with a well-developed formal geometry and geometrical astronomy.[38] And what are presently termed the occult sciences (or, not uncommonly, the pseudosciences)—astrology and magic— appeared in the East under the patronage of monarchs who valued them for their reputed utility both in predicting the future and in accomplishing economic miracles. In Greece, where no science of any kind was patronized by the city-states, the tradition of learned mysteries was eclipsed by natural philosophy, or was reformed to satisfy the interests of

TABLE 1. THE FRONTIER IN ANCIENT SCIENCE

Oriental Science (Mesopotamia and Egypt)	Hellenic Science (Ionia and Greece)
Anonymity (no biographies)	Scientific personalities (many biographies)
Bureaucratic civil services (court and temple patrons)	No employment
Institutions of learning (scribal schools; House of Life [Egypt])	No institutions
Useful knowledge (astronomy/astrology; applied mathematics; calendrical reckoning)	Abstract theory (natural philosophy)
"Science of lists" (extensive knowledge)	Logic; theoretical principles (intensive knowledge)
Occult learning strongly patronized	Occult learning eclipsed by natural philosophy
Scientific characteristics repeated independently by other hydraulic civilizations	Unique scientific culture in the ancient world

individuals, as when astrology in the Hellenistic world became "personal or horoscopic astrology" (Neugebauer, 1957: 100).

These are the characteristics of the scientific traditions of the ancient world that serve as markers along the frontier between Oriental and Greek science—the anonymity of the Oriental savants, their organization as a bureaucratic civil service, their network of learned institutions, the severely utilitarian objectives of their research, the high level of development of their calendrical reckoning, their corresponding indifference to any philosophy of nature, their compilatory method of treating knowledge, the independent recurrence of similar scientific cultures in widely-separated economies based on hydraulic agriculture, and their nurture of the occult sciences as a service to statecraft. Greek science stood apart along the whole front (Table 1). These contrasts delineate an intellectual and cultural frontier between East and West, a frontier that coincided with a geographical boundary.

It is this alignment between cultural developments and material conditions that can serve as the foundation for a theoretical geography of ancient science.

TOWARDS A BINARY HISTORY OF SCIENCE

Historical studies, unlike the natural sciences, admit no universal laws. A theoretical model may reflect, or rather refract, a social pattern. Or it may at least suggest one. In this bipolar model of the geography of science based on the scientific cultures of the ancient Oriental kingdoms and of Hellenic Greece, the characteristics of each pole are necessarily overdrawn. In the ancient kingdoms intellectual curiosity could have found niches in even the most absolutist bureaucracies. Conversely, in classical Greece studies that were aimed at useful results in the fields of medicine, geography, philology, and astrology were sometimes pursued without government support or intervention.

These qualifications do not, however, invalidate the model. It may remain suggestive and fertile within its limitations. Insofar as it is possible to cast the history of science in a binary mold shaped by geographical considerations, the possibility rests ultimately on the validity of the ecological analysis of ancient societies. The theory of "Oriental despotism," or, in its less flamboyant formulation, the hydraulic hypothesis, has sometimes been misconstrued, and repudiated accordingly, as a strictly determinist, monocausal explanation: big technology, necessitated by artificial irrigation, led to big government. Hence, whenever big technology is missing from the archaeological record or is not judged to have been big enough, the entire explanation is dismissed. A more plausible account locates hydraulic agriculture in a matrix of interdependent processes among which the construction and maintenance of the hydraulic infrastructure is a major component. Just how large a component it had to be to trigger social and political centralization has not been precisely determined, although it has been studied in specific societies (see, for example, Pasternak, 1972). Compared with modern industrial economies the economy of an ancient civilization, however seemingly opulent, was minuscule. In ancient Egypt, the population of the Nile Valley never exceeded 2,500,000. Power was supplied by animal and human muscle. It was only in the New Kingdom (1570–1070 B.C.) that the simple shadoof was first used to raise water, and the animal-powered water wheel was not introduced until the Hellenistic period (Butzer, 1976: 85, 46). At that level of technical development even modest engineering projects that were vital to the economy and that needed constant tinkering and attention could have required a sizable corvée and the intervention of an elaborate social organization.

Although Wittfogel might have considered it unwarranted and objectionable, his thesis may be separated into its ideological and its analytical components. Quite apart from the questionable accuracy of some of his judgments, his political program is perhaps best left to the dwindling cohort on the anticommunist ramparts. But the analytical implications of his ecological interpretation come into play in the study of social change and, specifically, have a bearing on the history and geography of science.

To the extent that the hydraulic hypothesis is valid, it belies the notion of a "unitary science" in the ancient world. Instead, science would appear to have had a double root—Oriental and Greek—the scientific tradition projected by the hydraulic societies of the ancient Middle East markedly different from the tradition that sprouted in Greece. This bipolar historical model is consistent with the opposition in all scientific cultures between pure and applied science and between the individual scientist's intellectual curiosity about nature and numbers (the Hellenic heritage) and society's efforts to direct research along utilitarian lines (the Oriental heritage). That parallelism is suggestive and meaningful. It should then be possible to identify the offshoots of these roots in their Hellenistic, medieval, and modern developments and to correlate them with the material and cultural characteristics of their respective habitats.

CHAPTER TWO

=

SCIENCE IN THE ASIATIC MODE

OF PRODUCTION

FROM STAGES TO ZONES

History focuses on development over time, geography on differences across space. These alternate perspectives can be complementary, but too often they are exclusive. The historical approach, divorced from geographical considerations, leaves the impression that societies, systems of governance, and modes of production occur as temporal sequences, sometimes progressive, but in any case connected in a linear process. It is commonly assumed that one stage of development gives way to its successor and that the succession occurs under pressure of political and economic events alone.

In general, however, modes of production and social systems do not develop in a single linear order, progressive or otherwise. They are separated by space and physical conditions as often as by time, class structure, and cultural development. They may occupy different geographical zones and may not, in any meaningful sense, constitute a sequence. In Mongolia and China, pastoral nomadism and settled agriculture developed (and continue to function) separately in adjacent but ecologically distinct zones—the former in a region too deprived of surface and atmospheric moisture to sustain agriculture on any terms, the latter in a region where intensified agriculture today supports a population of one billion people.[1] For millennia the two zones have remained as sociologically disjoint as they are ecologically dissimilar.

In Europe feudalism was indeed the predecessor of capitalism in that habitat, and it may reasonably be argued that there was a physically connected and perhaps even "progressive" development from the one to the other. But in no sense can a connected line of development be traced from feudalism back through Greco-Roman ("slave") society to the ancient civilizations of the Middle East. Feudalism was separated from the Asiatic Mode of Production in the Fertile Crescent by more than a thou-

35

sand miles, and in its origins by more than a thousand years, to say nothing of the consequential differences between a fertile, rain-watered plain and desertified river valleys.

Marxist historiography has undoubtedly contributed to the tradition of treating social development sequentially, through its notion that one society emerges, under the pressure of class antagonism, from the "womb" of its predecessor. Eric Hobsbawm has called attention to the confusion:

While these different forms of the social division of labour are clearly alternative forms of the break-up of communal society, they are apparently presented [by Marx] as *successive* historical stages. In the literal sense this is plainly untrue, for not only did the Asiatic mode of production co-exist with all the rest, but there is no suggestion in the argument of [Marx's] *Formen*, or anywhere else, that the ancient [i.e., Greco-Roman] mode evolved out of it. We ought therefore to understand Marx not as referring to chronological succession, or even to the evolution of one system out of its predecessor . . . but to evolution in a more general sense. (Hobsbawm, 1964: 36)

This "more general sense" is, however, hardly clarified by a discussion of the ways by which "each of these systems is in crucial respects further removed from the primitive state of man" (Hobsbawm, 1964: 38). A fuller clarification is reached by considering the material conditions, across spatial and geographical boundaries, on which these alternative forms of society were based.

Neolithic farming, in the form of small-scale village gardening without the use of irrigation or the plow, intervened between Paleolithic food collecting and urban civilization in a few regions of the world. Bureaucratically administered societies, with their impacted populations in constricted river valleys and their courts, palaces, and monuments, could not have sprung directly from conditions of Paleolithic food collecting without a transition through small-scale settled agriculture. Food-collecting nomads, once they were impelled into a settled way of life, sometimes progressed through a sequence of mixed economies; and only after a long process of development did some of them, under specific ecological conditions, construct agricultural systems with sufficient surpluses to support urban civilization.

In some habitats Neolithic gardening never evolved into an "Asiatic" economy, either remaining arrested or, depending on material conditions, evolving into some variant social system. The Neolithic settlements that formed at an early period in Northern Europe eventually coalesced into feudal concentrations and then, as their populations increased, into nation-states, without acquiring the earmarks of the Asiatic Mode of Production. The moist, clayey soil of the North European plain could not sustain dense and urbanized populations prior to the introduction of horse power and the heavy plow during the "agricultural revolution" of the Middle Ages (White, 1962a). When those technological innovations came into play, the agriculture of Europe, dependent on the rainfall that was beyond the control of any earthly potentate rather than on artificially distributed surface or ground water, then shaped the development of societies that at least in their early stages were sharply divergent from the grandiose and centralized civilizations of the arid, subtropical East.[2]

Analyses of societies as diverse as those of Western Europe and those that Wittfogel designated as "hydraulic" and Marx called "Asiatic" require that the historical stage be coupled with the geographical zone in order to construct a geography of history and ultimately a geography of science. Among the social systems that Marx postulated, the Asiatic Mode of Production provides an ideal principle for gathering and comparing empirical evidence and for defining zones in which geography and culture may be correlated, because it has appeared, spontaneously or derivatively, in widely separated regions of the Old and New worlds. It thus presents a variety of opportunities to measure both the similarities that collectively define a typical society and the contrasts that may help explain disparate cultural patterns. In order to test the plausibility of an ecology-society-science complex, four hydraulic societies, in different stages of development and located over a range of twelve thousand miles across the Old and New worlds, will be reviewed. Attention will be drawn, not to consensus among scholars (on issues that are inherently difficult to resolve), but rather to controversy and differences of opinion.

Each of the cases was selected for its exemplification of a different aspect of the argument. The history of ancient Ceylon provides an opportunity to review an influential challenge to Wittfogel's hydraulic hypothesis. The history of Chinese science and society, at least as developed by Needham, confirms the principles of the geography of science. Maya civilization appeared until recently to be a major exception to the claims of the hydraulic hypothesis; the recognition that it is typical rather than exceptional now lends authority to the geographical interpretation of ancient civilizations. Finally, the Pueblo cultures of the southwestern United States exhibit societies in an intermediate stage of Asiatic development in an ecological zone with limited hydraulic potential.

THE TEARDROP OF INDIA

Poised twenty miles off the southern tip of India, within 10° of the equator, lies the island of Ceylon (now Sri Lanka). Its self-contained compactness and its sharp division into a dry zone and a wet zone (the former, ironically, serving as the site of an ancient agricultural civilization) make it an ideal social laboratory for studying the connections between physical habitats and the societies they sustain.

The interaction of topography, hydrogeology, and climate determined the conditions under which Ceylonese civilization arose and developed. The southwestern quadrant of Ceylon is elevated, rising above five thousand feet. (See Figure 2.) During the spring, the southwest monsoon approaches the island, laden with moisture from the Indian Ocean. As it encounters the high ground it is deflected upwards into altitudes of lower temperature where its moisture is precipitated, producing the wet zone. With their moisture wrung from them, the winds then pass over the rest of the island, the dry zone, where they not only fail to produce rain but, instead, act as a desiccating agent (Farmer, 1957: 24). Then, during the winter months, from November to January, the northeast monsoon traverses the island. As it moves across the lowland dry zone, undeflected by any high ground, it produces only an erratic wet season; but as the still moist winds are forced upwards by the sharp relief of the southwestern part of the island, they once more drench the wet zone.

Today, most of Sri Lanka's farming, along with the bulk of its population, are situated in the wet zone. In ancient Ceylon, however, a sharply different social pattern developed, the wet zone supporting only low-intensity slash-and-burn farming while the dry zone sustained a high civilization based on intensive agriculture requiring elaborate irrigation systems.[3]

The differentiation of the two zones by rainfall patterns is based on both seasonal and annual levels. In the wet zone, where the mean annual rainfall ranges from seventy-five to more than two hundred inches, the region is sodden during both monsoon seasons, and throughout the year rainfall may be relied upon to maintain soil moisture at a level above the wilting point (Farmer, 1957: 25–27). The result of this precipitation system is a lush region where, since the demise of the ancient civilization, rain-watered agriculture has been well rewarded. Conversely, in the dry zone, which in the modern period has been sparsely populated and is now the scene of intense efforts at recolonization, rainfall, despite generally adequate mean annual levels, is variable and unreliable even in the wet season, since the southwest monsoon contributes little water and the northeast monsoon is fickle. Years of greater rainfall alternate with years of drought. During the dry zone's dry season, which unfortunately for farmers occurs in the summer when temperature and sunlight are ideal for agriculture, unirrigated farming is scarcely possible; and even in the wet season during the winter, lower temperatures, reduced sunlight, and frequent periods of insufficient precipitation make dryland farming discouragingly risky, especially in the case of a moisture-demanding plant

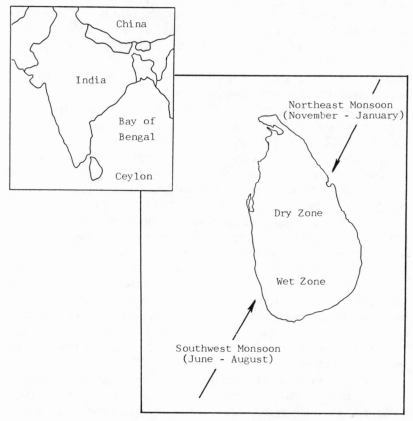

Fig. 2. Ceylon (now Sri Lanka). The island is separated from southern India by a narrow strait. Both monsoons produce wet seasons in the wet zone. In the dry zone the northeast monsoon produces only an erratic wet season, while the southwest monsoon actually intensifies the aridity of the zone.

like rice (which germinates under water). Nonetheless, in a striking social and geographical paradox, it was in the dry zone that the ancient Sinhalese (the Ceylonese branch of Buddhists) and the Tamils created, and maintained for fifteen hundred years, a splendid civilization that displayed the characteristics of the Asiatic Mode of Production.[4]

Ancient chronicles, archaeological research, documentary evidence, and tradition all point to Ceylon as the setting of a high civilization based on irrigation agriculture. Long before the Bengal invasion of the island (sixth century B.C.), and even before the hydraulic farming that the Yakkha aborigines are believed to have initiated, came the mythical King Ravana, reputed to have been a water king, an "expert engineer" (Fernando, 1982: 43). The very first sovereigns who invaded Ceylon from the Indian mainland, twenty-six hundred years ago, bent their efforts to the development of agriculture in the dry zone. James Emerson Tennent, the nineteenth-century historian of Ceylon who served as British colonial secretary on the island from 1845 to 1850, insisted that Wijayo, the founder of the first dynasty, was a water king and not a religious reformer: "Whatever may have been his first intentions, his subsequent policy was rather that of an agriculturist than an apostle. . . . [The] earliest attention of the Bengal conquerors was directed to the introduction and extension of agriculture" (Tennent, 1977, vol. 1: 288–89).

It may be that those invaders arrived from a region of eastern India with topographical and atmospheric features similar to those they found in the dry zone and, hence, simply applied an agricultural technology with which they were familiar. Or, regardless of prior experience, they may have recognized the long-term benefits of establishing irrigation agriculture in a potentially fertile dry region, despite the high infrastructural start-up costs of irrigation systems. Conversely, they may have found the efforts required to clear forests and perhaps contend with pests, pestilence, and predators in order to establish rainfall farming in the wet zone comparatively unrewarding. Moreover, agrogeographical conditions made the dry zone better adapted for production of the grain crops that have always served as staples in agricultural societies. Even today, the wet zone is less suitable for the production of these staples than for the production of export crops, including the famous Ceylon teas.

In any event, the first two centuries after the conquest were devoted to constructing the basis of a hydraulic civilization. In a land where surface water was sparse, the techniques of river-valley agriculture were of limited benefit. Instead, an irrigation technology was developed that utilized and sometimes combined a variety of technical systems, including thousands of earthen tanks (reservoirs) designed to catch the runoff of the irregular rainfall. The earliest public work recorded in the ancient chronicles was a tank constructed by the successor of the first Bengal king (Tennent, 1977, vol. 1: 290); and during succeeding centuries, many thousands of additional tanks were built.

Minor hydraulic structures, including small canals and tanks, which might be the work of a village or even a single farmer, were located mainly in the wet zone, where no centralized authority evolved and no high civilization was created during the ancient period. But major hydraulic systems, including large canals, massive dams, and complex arrangements that combined tanks designed to catch runoff and feeder canals that added water from nearby streams to the charge of the tanks, were exclusively confined to the dry zone, which became the habitat of the great Sinhalese civilizations (Fernando, 1982: 44). Joseph Needham has sung the praises of the extraordinary skill of the engineers of ancient Ceylon:

It will be evident even from this roughest of sketches that the achievements of Indian civil engineers in ancient and medieval times are quite worthy to be compared with those of their Chinese colleagues, though not to win the palm. Yet it was never in India that the fusion of the Egyptian and Babylonian patterns achieved its most complete and subtlest form. This took place in Ceylon, the work of both cultures, Sinhalese and Tamil, but especially the former. (Needham, 1971a: 368)

According to most authorities, these complex and large-scale projects could have been organized only as the public works of a central government. M. B. Ariyapala, former president of the Royal Asiatic Society (Sri Lanka Branch), who has been described as a "highly expert" authority (Leach, 1959: 22), stated that "the supplying of water to the fields was too heavy a task for the cultivator, and therefore the central government had to undertake the responsibility of providing the necessary water. . . . The long history of the island records the efforts of many kings to improve the irrigation schemes of the island as an aid to agricultural activity." Moreover, the "extensive irrigation works needed much attention and a great deal of labour to keep them in repair" (Ariyapala, 1956: 329–30). The labor required to build and maintain these public works was exacted by the kings through a system of compulsory conscription known as *Rajakariya*, the ancient Ceylonese form of the corvée. And because the physical conditions of irrigation agriculture were heavily dependent on such communal efforts and produced consequences that frequently spanned several villages or even an entire region, it often fell to the central authority, the sovereigns, to allocate the flow of water and to fix the boundaries of paddy and cropland (Tennent, 1977, vol. 1: 364, 289). There can be no doubt that the Sinhalese system of hydraulic agriculture produced enormous agricultural surpluses and large urban population concentrations; at its peak, in the twelfth century, the capital, Polonnaruwa, on the northeast coast of Ceylon, may have been the most populous city in the world (Chandler & Fox, 1974: 363).

The agrotechnology of ancient Ceylon displayed the unmistakable features of hydraulic agriculture, and the cultural accomplishments and institutional superstructure that rose from this technoeconomic base were equally typical. In addition to the irrigation system, the elaborate masonry architecture included shrines, temples, and palaces, some of which "are on much the same scale as the Pyramids of Egypt—there are at least two which contain over twenty million cubic feet of solid brickwork" (Leach, 1959: 12).[5]

Although the history of Ceylonese science awaits its Needham, there are indications in it of a typically Oriental pattern of scientific organization and development. Accomplishments in astronomy, astrology, arithmetic, medicine, alchemy, geology, and acoustics are all referred to in the ancient chronicles; and the patronage of kings and the participation of temple scholars are mentioned, indicating that science was bureaucratically organized. Medicine and public health have been chronicled with particular attention, and there are inscriptions showing that "the chief physician . . . was one of the principal functionaries of the State." Public health "was no doubt one of the chief concerns of the rulers of ancient Lanka, and they did much to promote it," by the establishment and support of health-related public institutions—hospitals, lying-in homes, dispensaries, kitchens, and medicine-halls (Ariyapala, 1956: 273–74, 278–79, 281).

These achievements of Ceylonese society are direct testimony, but there is indirect evidence that lends further support to the impression that the social organization of Ceylonese science was similar to the scientific traditions of the hydraulic kingdoms of Egypt and Mesopotamia. At Angkor Wat, in Cambodia, whose ancient history parallels that of Ceylon, and whose irrigation system was "the most impressive . . . in Southeast Asian history" (Bray, 1986: 72), astronomy, calendrical reckoning, astrology, religion, and numerology were held to form a "unity" and were patronized as temple studies by priest-architects.[6] The temples themselves were oriented and constructed along lunar, solar, and solstitial alignments; and in some structures astronomy, astrology, numerology, and architecture were fully integrated (Stencel, Gifford, & Moron, 1976). Moreover, the astronomers of these Southeast Asian civilizations, particularly the Tamil astronomers of Ceylon, displayed an affinity with the bureaucratized sciences, including astrology, of the ancient Middle East at the same time that they remained indifferent to the abstractions of the classical Greeks (Neugebauer, 1952: 252–54).[7]

Despite these indications that the case of Ceylon confirms the hydraulic hypothesis in many details, its history has been seized upon in criticizing Wittfogel's thesis. Given what appears to be a typical hydraulic civilization, Wittfogel might have been expected to cite Ceylonese history as a compelling confirmation of his views. But, curiously, he neglected it entirely in his major study, Oriental Despotism, in which he discussed many examples of "agromanagerial" societies. Conversely, in a peculiar historiographical inversion, E. R. Leach, who was an eminent anthropologist and whose analysis of Wittfogel's interpretations has been described as a "trenchant critique," conceded that "a recent general survey . . . of the various types of available evidence for this region [dry zone Ceylon]—archaeological, epigraphic, documentary and ecological—has drawn attention to the perfect manner in which Ceylon fits the requirements of Wittfogel's thesis" (Leach, 1959: 6).

Far from trenchant, Leach's criticism of Wittfogel was tentative, ambiguous, conjectural, and compromising. He admitted "the puzzling historical fact that many of the most successful archaic societies were located in flood-ridden arid terrain in which agriculture was only made possible through the use of irrigation engineering"; he accepted the term "hydraulic societies" as a valid description of these archaic social formations; he acknowledged that "the great kings have reputations as irrigation engineers rather than as conquerors or as builders of cities," and that the "major works are numerous and immense in scale"; he called attention to the institutionalized labor corvée "known as *rajakariya*, i.e. king's work"; and he credited Wittfogel with isolating "the critical features" of these ancient societies: "Despite my critical comments I hold that Wittfogel in his concept of 'the hydraulic based despotism' has picked out for emphasis the critical features of an important type of archaic civilisation which is well represented by the Sinhala type" (Leach, 1959: 2, 5, 8, 10, 14, 16). But he denied—and this was his central objection to Wittfogel's thesis—that these hydraulic conditions can be correlated with highly centralized bureaucratic authority.

Leach's argument, which is potentially damaging to the hydraulic hypothesis only in its more intemperate formulations, is inconclusive mainly because the evidence he marshalled is incomplete, tendentious, and conjectural. He accepted the testimony of a seventeenth-century European observer who had been a long-time resident of the Kandy kingdom of central Ceylon—"Robert Knox's outstandingly perceptive description of this Kandyan kingdom . . . based on a twenty-year period of residence." When, however, Knox's account confirmed the king's role in deploying a teeming corvée on "vast works," Leach repudiated it with a gratuitous attribution of unconscious bias:

Is not he too, like Wittfogel, engaged in political polemic? . . . [We] need to remember that he himself was a citizen of seventeenth-century England, a country in which the divine right of despotic kings was still a prime political issue. Knox's detailed elaboration of the Kandyan king's tyranny is perhaps, in part at least, a slanderous attack on monarchy in general. (Leach, 1959: 11, 12)

Leach also questioned the whole Sinhalese tradition and its extant chronicles on equally skeptical and conjectural grounds:

Can we really infer from a tradition of "great" kings that the kings in question were truly great? Might it not be that the "greatness" of the hydraulic monarch is itself a product of propaganda myth? Need we believe the Sinhalese chronicles any more than we believe Knox? (Leach, 1959: 13)

But it is not the chronicles alone that have convinced most authorities that the ancient hydraulic works were vast and ingenious and required royal patronage and direction. Archaeological evidence, which is still being uncovered as the redevelopment of the dry zone proceeds, consistently buttresses the traditional accounts (Fernando, 1980). Even if the many ancient records of kings who applied themselves to irrigation and various water-control projects are the result of a "propaganda myth" (and there is no evidence for it), the fact alone that such importance is attached to legends of water kings, in Ceylon as well as in other hydraulic societies, supports the inference that hydraulic engineering was an activity of considerable social importance and of central interest to the government authorities. Moreover, the chronicles attest not only to the greatness of the kings but also to their decline in step with the decline of Ceylonese agriculture in its later period: "The *Rajavaliya* . . . records: 'because the fertility of the land was decreased, the kings who followed were no longer of such consequence as those who went before'" (Brohier, 1975: 18). Is this failure, as well as the success, a product of the same "propaganda myth"?

Leach, whose influential article is inconsistent with the conclusions of most authorities, suggested that not only were the kings not great but the hydraulic engineering of ancient Ceylon was on the whole comparably inconsequential, requiring no great efforts and hence no strong centralized organizational authority. After granting that "the Ceylon irrigation works and the associated palace and temple construction works do represent a gigantic accumulation of capital resulting from an enormous number of man-hours of labour," he called attention to the long period of time—fourteen hundred years—over which these colossal works were built, and he drew the conclusion that "their construction was haphazard and discontinuous" (Leach, 1959: 23). This argument, along with the circumstance that the kings sometimes granted tracts of land to their followers, led Leach to invent a historiographical oddity in order to characterize ancient Ceylonese society: nonmilitary "hydraulic-oriental feudalism" (Leach, 1959: 17).

Leach's argument is effective against, and indeed is directed only against, an extremist formulation of the theory of Oriental despotism; but it is distinctly unproven and unconvincing against more temperate interpretations of the hydraulic hypothesis, including some of those put forward by Wittfogel himself. The target of Leach's criticism is an untenable and unnecessarily extreme statement of the conditions that sustain hydraulic societies and produce bureaucratic intervention. Referring to one of Wittfogel's (less typical) assertions, Leach remarked:

Note the assumptions involved, particularly those implied in the term "full aridity." Wittfogel's thesis is not simply that areas exist in which cultivation is

only possible if they are artificially irrigated, but rather that his type system—
"hydraulic society"—grows up where cultivation calls for *large scale* irrigation;
the initial hypothesis is that in such societies *all* water supplies must be brought
from afar. . . . But if we ask the question: "Is it an historical fact that most of the
hydraulic civilisations of the past have grown up in 'fully arid' regions?", the
answer is "No!" Dry certainly, but not "fully arid." (Leach, 1959: 7)

Quite apart from the questionable accuracy of this quibbling characterization of Wittfogel's essential idea, it should be obvious that in each
society the magnitude of the factors must inevitably vary. A society may
be hydraulic to a greater or lesser degree, just as in any taxonomic system
a taxon may have more or fewer characteristic traits without necessarily
requiring reclassification and without discrediting the system of classification itself. The hydraulic hypothesis is not a question of all or nothing.
A more reasonable question is whether there is sufficient correlation
between greater and lesser hydraulic engineering projects and corresponding intensities of political centralization, a question that calls for
comparative analysis across a variety of hydraulic societies. A tentative
answer, favorable to the hydraulic hypothesis and cognizant of Ceylon's
complex meteorological and hydrological conditions, has been offered by
Needham:

The thought lies near that if that intimate relation suspected by so many between
feudal-bureaucratism [Needham's euphemism for the Asiatic Mode of Produc-
tion] and hydraulic works should prove to be well-founded, the precise character
of the hydraulic works may have great sociological importance. If Ceylon seems
not to have generated that form of society very markedly, while China undoubtedly did, the differences considered in this Section [on hydraulics] may well be
meaningful relative to this diversity. (Needham, 1971a: 375)

Leach's conclusions may be questioned further, insofar as they depend
on his assumption that the Ceylonese irrigation network was constructed in so piecemeal a manner that no centralization of authority was
called for. In fact, the engineering works in question consisted not only of
a variety of massive and complex structures but also of many thousands
of tanks distributed over the dry zone. At a uniform rate of construction, a
tank would have been built every few weeks over a fourteen hundred–
year period. By a more realistic measurement, there would have been
periods of intense construction alternating with periods when only
maintenance and normal operation were required. If modern examples,
such as the Tennessee Valley Authority, can serve as a guide, the centralization that initial construction entails persists long after the work is
completed, as ownership, operation, regulation, maintenance, and extension remain the province of the central government. Besides, the construction of irrigation systems was only part of a water king's duties. On a

more continuous basis, even beyond the responsibilities of permanent superintendence of the infrastructure, he assigned rights to water impounded and distributed by catchments and canals serving entire districts, and he arranged for the storage, security, and distribution of the surplus crops that intensive agriculture made possible.[8] The belief that the elaborate Sinhalese hydraulic system produced no centralization of authority suggests an unfamiliarity with the world of large-scale engineering and its interconnections with government. Leach's skepticism about the political consequences of Ceylonese irrigation engineering is as unwarranted as the opposite assumption to which he objects—that in the presence of hydraulic agriculture there must inevitably be a completely centralized and despotic authority.[9]

It may be conceded that in the course of over a millennium the Asiatic Mode of Production in ancient Ceylon was intermittently diluted by decentralized authority patterns and its intensity diffused by the appearance of "feudal" enclaves, just as European feudalism was in its turn adulterated by nonfeudal arrangements. In Europe, feudalism sometimes functioned on a quite subordinate level, and even in pockets not at all. Marc Bloch observed that "feudal Europe was not all feudalized in the same degree or according to the same rhythm and, above all, that it was nowhere feudalized completely" (Bloch, 1961, vol. 2: 445). Indeed, Sweden was never feudal, and Spain only belatedly. Neither Bloch nor any other historian would consider those circumstances grounds for denying that Europe was feudal. Ancient Ceylon was not an extreme case of Oriental despotism, but in the opinion of almost every authority the evidence confirms the view that a bureaucratic state played a central role in the management of its hydraulic agriculture.[10]

The form of the hydraulic hypothesis that is usually favored by its advocates accounts for conditions where centralization is less than total while it retains the principle that hydraulic agriculture is a centralizing force that, depending on topography, climate, hydrogeological conditions, and the size of the economy compared to the scope of the construction work, produces corresponding degrees of centralization. To the limited extent that Leach's objections are valid, it appears that the geography of dry zone Ceylon, with its lack of sufficient surface water to permit many large-scale impounding and channeling projects and its encouragement, instead, of numerous small tanks to retain runoff, fostered a less intensive centralization of government than the paradigmatic type of Asiatic society that Wittfogel has generally chosen to study.

Whatever the exact circumstances, there was sufficient concentration of authority to oversee the construction, maintenance, and control of complex projects in an extensive network of large and small hydraulic works, some "equipped with scientifically designed spillways and sluices" (Leach, 1959: 8), to create architecturally impressive shrines, temples, and palaces, and to authorize and patronize a typical Oriental pattern of court and temple science.

YÜ, THE WATER KING

In a reassuring display of the unity of culture and material life, hydraulic societies have often preserved hydraulic legends, some of them pertaining to great feats of engineering performed by founding monarchs. There were Wijayo and King Ravana in Ceylon. In Egypt, the first Pharaoh, Menes, is reputed to have embanked the Nile at Thebes. And in China, Yü the Great, the semilegendary founder of the first dynasty (Hsia), was "the first ruler to undertake conservancy works on a large scale" (Lattimore, 1951: 33–34). His exploits have been commemorated by word and woodcut (Needham, 1971a: 213, 233; 1984: 147). "Yü the Great controlled the waters—such is the phrase which has become proverbial, and which one encounters all over China. Probably no other people in the world have preserved a mass of legendary material into which it is so clearly possible to trace back the engineering problems of remote times" (Needham, 1971a: 247).

Although these accounts of builders of irrigation systems, flood control projects, and transportation canals are undoubtedly blends of legend and history, sinologists have attached considerable importance to them. Needham lent his authority to the opinion that "the Yü legend incorporates two basic features of Chinese agrarian society, the regulation of waterways and the use of organized corvee labour for accomplishing it" (Needham, 1971a: 251).

However, much more so than in the study of Ceylonese history, interpretations vary widely about the nature of the social organization that was based on the hydraulic infrastructure, and a baffling variety of designations have been favored and repudiated in classifying traditional Chinese society. The terms feudal and Asiatic, which in many ways represent polar opposites (the former highly decentralized and lacking urbanization and public works), have been applied with equal fervor. Among Marxist writers the term Asiatic Mode of Production long remained taboo (Schwartz, 1953), and it has no appeal to those non-Marxists who choose in general to deemphasize the importance of environmental factors in sociohistorical analysis. At one point Wittfogel favored a feudal rather than Asiatic interpretation of early Chinese society and used the term Oriental feudalism to describe it. Upon reconsideration ("intensified comparative studies") he revised his diagnosis and referred to the institutions of early (Chou) Chinese monarchs as "those of a hydraulic society, which gradually intensified its managerial and bureaucratic 'density'" (Wittfogel, 1981b: 33n). His rejection of a feudal interpretation was based largely on the contrast between the great extent of hydraulic public works in China and the absence of large-scale public works of any kind in prototypical feudal Europe. Owen Lattimore, also referring to ancient China, postulated a transition, which he described as a "feudal phase of organization," between incipient farming and "a bureaucratically administered empire"; but he, too, recognized the analytical difficulties, and conceded that any postulated Chinese feudalism would have been a strange society indeed:

[Chinese feudalism] was necessarily interested in public works of a kind that were not characteristic of European feudalism. The labor of serfs on the domain of the [European] feudal lord did not meet all the requirements of an economy founded [as in China] on irrigation. In addition, the labor of the whole community had to be regulated for the maintenance of public works—water rights had to be allotted and grain stored and issued for the labor gangs. (Lattimore, 1951: 375)

Some historians have employed the terms "feudal" and "feudalistic" only provisionally in characterizing ancient China. Rushton Coulborn wrote that "Chou China . . . may in fact have become feudal but the final proof is lacking" (Coulborn, 1965: 215). Needham prefers never to use *Asiatic Mode of Production* as a designation (the terminology may offend his internationalist sensibilities), and postulates instead a direct transition from "primitive collectivism" to "proto-feudalism" and then to "feudal bureaucratism" (Needham, 1971a: 251).

Wolfram Eberhard discarded all interpretations that emphasize economic conditions as excessively materialistic and proposed instead that agrarian ancient China was "super-stratified" by pastoral invaders from the West (Eberhard, 1965: chs. 1–3). The result, according to Eberhard, was a nonhydraulic "gentry society" with feudal traits that were the product of "certain social changes [in which] power played a far greater role than economy" (Eberhard, 1965: 58). He dismissed all evidence of an early development of irrigation agriculture along the Yellow River; he found Wittfogel's formulations inconsistent and extravagant; and he denied that Chinese society was markedly centralized or despotic. Nonetheless, he conceded that the provincial governors and magistrates were appointed by the central government "and subject to change every three years or less" (Eberhard, 1965: 63–64). And, at least with regard to the cultivation of rice in South China, he stated: "There is no doubt that there is a correlation between irrigation and rice. . . . Further research may discover here a whole complex with irrigation as the central element, influencing all other elements. It seems indeed to be useful to study this complex in detail" (Eberhard, 1965: 88).

Derk Bodde attempted to reconcile these multiply conflicting claims.

He agreed with Eberhard that "Wittfogel has probably considerably over-estimated the all-embracing role of irrigation in China, at least as far as the North is concerned" (Bodde, 1965: 79). But he questioned on linguistic grounds Eberhard's conjecture of an invasion of Turkish pastoralists; he found it hard to believe that the farmers of arid and semi-arid North China would not have developed an irrigation technology at all; and he called attention (as, indeed, Wittfogel himself often did) to the other hydraulic enterprises in China—drainage, flood control, and canal transportation—which, on a sufficiently large scale, would have produced the same kind of centralized institutions as irrigation engineering (Bodde, 1965: 80–81).

The taxonomy of social systems (like all taxonomies) inevitably displays some ambiguity as well as considerable overlap and interpenetration of traits among taxa meant to be disjoint. The ancient Near East and classical Greece were relatively neat categories separated by comparatively hard-edged characteristics. The former, compact and geographically homogeneous, was intensely hydraulicized but failed to achieve an Iron Age, while Greece depended on a rain-watered agricultural base supplemented locally by irrigation techniques, an iron technology, and extensive trade and commerce. By contrast, in China, with its vast extent and varied geography, the taxonomic uncertainties were correspondingly more acute. A hydraulic agriculture whose productivity was increased by the introduction of iron implements (600–500 B.C.) was sometimes combined with private ownership of land and a feudal type of social stratification. Moreover, with regard to China and the materialist interpretation of its history, analysis is confused by politics to an extreme degree. Even Needham, in the course of his studies of Chinese science, expressed uncertainty about the grounds on which the concept of the Asiatic Mode of Production had been suppressed "by those [Marxists] who opposed any variation from the standard succession" and who "gained the day." Writing in the Marxist journal, *Science and Society*, he commented that "no doubt the climate of dogmatism which prevailed in the social sciences during the personality-cult period played some role in this situation" (Needham, 1969a: 203).

On the question of classifying the civilization of ancient China, Needham agreed with Wittfogel's theory of hydraulic society without, however, going so far as to adopt the designation of the Asiatic Mode of Production, much less the theory of Oriental despotism:

Although Wittfogel has perhaps overdone it, I do not regard his theory of "hydraulic society" as essentially erroneous, for I also believe that the spatial range of public works (river control, irrigation and the building of transport canals) in Chinese history transcended time after time the barriers between the territories of individual feudal or protofeudal lords. It thus invariably tended to concentrate power at the centre. (Needham, 1969a: 204)

Despite his acknowledgment of the determining influence that ecological conditions have had on Chinese society and agricultural technology and his admission that *feudalism*, because of its promiscuous overuse, "has become a meaningless term," Needham, presumably in order not to appear "unfriendly," has continued to prefer some compound designation like *feudal bureaucratism* in characterizing ancient Chinese society, while turning a blind eye to the Marxologically canonical but provocative *Asiatic Mode of Production*. His ideological and sentimental obligations have placed him in the predicament that typically ensnares those who attempt to harmonize a system of thought and analysis with a heartfelt soteriology.

A more poignant case of being impaled on the horns of ideological conviction and social analysis was that of the Chinese Marxist scholar, Chi Ch'ao-ting. In 1936 his study *Key Economic Areas in Chinese History as Revealed in the Development of Public Works for Water-Control* appeared in English, published in London. Chi followed Wittfogel's lead and built a case for interpreting Chinese history in terms of "key economic areas," each heavily hydraulicized and shifting their locations with changing economic conditions and strategic requirements. The "key" region, in which irrigation agriculture was the chief economic activity, served as "an economic base for the unified control of a group of more or less independent self-sufficient regional territories" (Chi, 1963: 149). Thus, despite the extent and diversity of Chinese geography, regions of hydraulic farming and large-scale water-control projects dominated surrounding regions where rainfall farming would have resulted in decentralized social and political structures. Chi's China was thus the inverse of Holland, where the institutions associated with dryland farming dominated the coastal districts where a form of hydraulic agriculture has been practiced.

Chi, like Needham, deferred to his ideological commitments and described China as feudal rather than Asiatic (except once, in quoting Marx). Later, as a member of the communist government in China, he evidently abandoned the hydraulic hypothesis altogether (Wittfogel, 1981b: 410). Only in 1979 was his book finally translated into Chinese and published in China, although its premises remain stigmatized (Rowe, 1985: 265).

Chinese science, while not entirely spared the obfuscation that has beclouded Chinese social and economic history, has been substantially secured by Needham's extensive studies. He described the "basic quality of Chinese astronomy" as "its official character and intimate connection with the government and bureaucracy" (Needham, 1959: 186). Its primary functions were calendrical reckoning in the service of agriculture and prognostication in the service of the emperor. It was administered by the Astronomical Bureau, whose principal functions were "research in calendrical science, compilation and distribution of calendars and ephemerides, and unremitting observation of astrological omens" (Yabuuti, 1973: 94–95).[11] Dynasties came and went but the Astronomical Bureau, which may originally have been linked to the development of hydraulic agriculture, lasted two thousand years and provided the institutional setting for Chinese astronomers, who were "on the whole bureaucrats first . . . and only secondarily researchers" (Yabuuti, 1973: 95–97).

Like Mesopotamian astronomy, but unlike that of the Greeks, Chinese astronomy applied only arithmetic and algebra to the study of celestial phenomena: "Unlike Greek astronomy, which employed geocentric models based on a well-developed Euclidean geometry, the Chinese, who were able arithmeticians and algebraists, succeeded in finding regularities in celestial phenomena without recourse to geometrical models" (Yabuuti, 1973: 96). Also, again as in Mesopotamia, the organization of knowledge through the making of lists of "natural products—probably minerals, chemicals and plants"—was a feature of early Chinese science (Ronan, 1978: 145).[12]

These are tell-tale marks of science in the Asiatic Mode of Production; but Wolfram Eberhard, commenting as a social historian, again cast a dissenting vote. Since evil portents reflected defective statesmanship, astronomers were sometimes in a position to play a political role by publicizing or withholding information about royal mismanagement. Sinologists have called attention to the political use to which inauspicious astronomical and other observations were put by Chinese functionaries (Bielenstein, 1950), but Eberhard chose to emphasize it exclusively while denying that calendrical or any other economic interests came into play at all in shaping the organization of Chinese science. He insisted that Chinese astronomy was strictly "applied," but that it was applied only to politics and not to calendrical science bearing on agriculture in any way. He polarized his analysis to an extreme degree, concluding that "the specialists had no real interest in the development and refinement of scientific laws; they were interested only in 'applied political science,' i.e., applied not in technology but in politics only. . . . The function of astronomy, astrology, and meteorology . . . was purely political" (Eberhard, 1957: 70).

Even if Eberhard's interpretation of the history of ancient Chinese science were correct it would still indicate a bureaucratic organization and a utilitarian dedication to governmental interests. However, its exclusive emphasis on political objectives and denial of any economic applications is generally inconsistent with the findings of the history of science. Nathan Sivin, an authority on the history of Chinese science, has taken exception to Eberhard's conclusions, which he found to be based on "characterizations of both Chinese and Western astronomy greatly at variance with those noted by historians of science." He objected specifically to Eberhard's sharp polarization of factors that "the sources seem to consider complementary," and he stated that "what is most needed [is] an exploration of the balance between them and how it was maintained" (Sivin, 1969: 53).

In the face of these controversies over the social organization and culture of ancient China, to what extent does the history of Chinese science and society support a geographical interpretation? The Asiatic Mode of Production, Oriental feudalism, superstratification, feudal bureaucratism, proto-feudalism, Oriental despotism, gentry society—not even a prophet could learn the whole meaning of a society so obscured by variety and complexity, stretched in time over millennia and in space over vast and sharply diversified environments. Perhaps Needham has earned the right to serve as magistrate, at least for a provisional judgment, on the issues of Chinese astronomy, the environmental hypothesis, and the possibility of a history of science the binary elements of which are Oriental and Greek:

> For an agricultural economy, astronomical knowledge as regulator of the calendar was of prime importance. He who could give a calendar to the people would become their leader. More especially was this true, says Wittfogel, for an agricultural economy which depended so largely upon artificial irrigation; it was necessary to be forewarned of the melting of the snows and the consequent rise and fall of the rivers and their derivative canals, as well as of the beginning and end of the rainy monsoon season. In ancient and medieval China the promulgation of the calendar by the emperor was a right corresponding to the issue of minted coins, with image and superscription, by Western rulers. The use of it signified recognition of imperial authority. There can be little doubt that very concrete social reasons may be found for the official and governmental character of Chinese, in contrast to Greek, astronomy. (Needham, 1959: 189)

PULLTROUSER SWAMP

Lying oceans away from Engels' "great stretches of desert which extend from the Sahara . . . to the highest Asiatic plateau," the New World provides a setting where the environmental hypothesis of science and society may be tested in relative isolation from the centers of hydraulic civilization in the Near and Far East. Ceylon, although an island, was easily within reach of the continental civilizations of India, China, and Southeast Asia; and China, while it evidently developed Neolithic and then intensive agriculture on its own, has since been both a source and a recipient of ideas and techniques diffused across its western borders. By comparison, the Americas were almost completely insular until the sixteenth century. Whatever transoceanic influences they may have received before the European invasions would have been relatively slight, and probably insignificant in accounting for cultural traits over the prior two thousand years, a period during which some native American societies were reaching high levels of development. Hence, any similarities that may be detected between the New and Old worlds would be largely the result of parallel cultural evolution, and the possibility would present itself of searching for similar cultural adaptations to similar environmental conditions.

Did, then, an Asiatic system, parallel to the Old World's, appear in the Western Hemisphere in regions that could for ecological reasons sustain hydraulic society? And did it sponsor a bureaucratic astronomy dedicated to the needs of intensive agriculture and an astrology in the service of governance? Or, conversely, in this isolated hemisphere that lacked the plow, the wheel, the ox, and the horse, did the development of agrarian societies depart from the pattern predicted by the hydraulic hypothesis?

The arid and semi-arid conditions that might be expected to foster the rise of hydraulic societies occur in what is now the western territories of the United States, in parts of Central America, and along the western coasts and highlands of South America. The most promising conditions lie south of the Rio Grande, and, as is well known, these regions of Central and South America have been the homelands of a series of high civilizations, most notably the Maya, the Inca, and the Aztec.

The Maya, who reached "the highest degree of achievement in the field of astronomy within the Americas" (Aveni, 1980: 136) and who are credited with "the most sophisticated development of mathematics indigenous to the New World" (Closs, 1986a: 291), present the most instructive case study. Their ancestral habitat lies in parts of what are now Guatemala, Belize, Honduras, El Salvador, and the Yucatan Peninsula in Mexico. Only the northwestern corner of this area is semi-arid; the rest is mainly wet, jungle-covered lowlands. The dead cities of the Maya, in contrast with the desiccated ruins of the Near East, are buried in sodden overgrowth to such an extent that they have been difficult to find. As a result, many of the fundamental secrets of the physical adaptation and social organization of their inhabitants are only belatedly coming to light.

Although there is a dry season even in the wetlands, the early Maya farmer was faced with the problem of too much water rather than too little. This circumstance has confused Maya studies, and only recently has substantial evidence been uncovered that, among the Maya, excessive moisture evoked a different technology but the same social responses that insufficient moisture produced in the agricultural societies of the Old World. Too little water or too much, an agricultural society is challenged to respond to severe hydrological conditions with infrastructural initiatives that require large investments of labor, and hence favor centralized authority. As long as it was believed that the ancient Maya economy was based on the same kind of slash-and-burn agriculture that the descendents of the Maya practice today, the high civilization and advanced astronomy of the ancients appeared to be completely anomalous. The Maya were perceived to be unique among ancient civilizations in basing a state-level, complex society on extensive (that is, non-hydraulic), rather than intensive, agriculture. Recent discoveries have revised and normalized Maya history.

Maya numeration and reckoning began to develop twenty-five hundred years ago and came to be based on a vigesimal system modified for calendrical purposes, since a pure vigesimal calendar would give 400 (i.e., 20 × 20) days, a poor approximation of the chronological year (Lounsbury, 1978: 760). It employed various symbols to represent zero, and according to Anthony Aveni, an authority on native American astronomy, it became the most refined system of numeration in the world (Aveni, 1980: 136–37).[13]

From its origin, Maya mathematics was applied to various governmental and bureaucratic operations, "in records of trade, levies of tribute, mensuration, census, and other functions of government and religion," including especially "chronology and regulation of the calendar" (Lounsbury, 1978: 764). The written calendar that the Maya devised incorporated a 365-day solar year, lunar and Venutian cycles, eclipse calculations, and a curious 260-day "sacred almanac." Moreover, the "court arithmeticians and calendar keepers" were, like their Oriental counterparts and quite unlike Greek natural philosophers, nameless civil servants—"the contributors to the Maya 'sciences' remain totally anonymous" (Lounsbury, 1978: 760, 804).

The objectives of Maya research were strictly utilitarian, again like Oriental traditions, without a trace of the Hellenic passion "to find an etiology" simply for the pleasure of the search and the discovery. Michael D. Coe, an anthropologist who has studied the astronomical systems of ancient American societies, noted the similarity between Maya science and that of the Old World Orient and its dissimilarity from that of the Greeks: "Numerology ruled supreme in Mesoamerica, allying their astronomy much more closely with that of Mesopotamia than with the Greeks, whose obsession was geometry" (Coe, 1975: 30).

Calendrical knowledge developed characteristically around economic requirements, in the form of determining agricultural sequences, and around a practical interest in prognostication (Aveni, 1982a: 2–3). The latter trait of Maya astronomy is easily confirmed and has been frequently noted. J. E. S. Thompson, who was a highly regarded authority on Maya culture, stated that: "One must appreciate the impact of the divinatory aspects of the 260-day sacred almanac, and never lose sight of the fact that the ends of Maya astronomy were not scientific, but astrological. The Maya were interested in the heliacal rising of Venus because then the world was in danger" (Thompson, 1974: 97; see also Katz, 1972: 65). On both counts, economic and political, the court and temple patrons of Maya science, like their Old World compeers, evidently saw to it that the savants in their employ confined their interests to what was then regarded as useful knowledge.

An additional source of information and evidence bearing on the possible use of Maya astronomy as an adaptive strategy is Maya architecture. Until recently, the odd shapes, apparent misalignments, asymmetry, nonrectangularity, and nearly but not quite perpendicular intersections of Maya buildings defied understanding and suggested that the builders had been careless or not fully competent (see Aveni, 1982a: 11). But close study by astronomically literate historians, anthropologists, and archaeologists has now revealed that many of these seeming imperfections are in fact codifications of astronomical knowledge—"architecture was for the Maya what pen and parchment were for the Greeks." What had seemed to be slipshod construction is now seen as a technique for establishing astronomical lines of sight in connection with economic activities. In the temple of Venus, at Copan in Belize, the view through a window that appears to have been designed for observational purposes reveals "an intriguing set of Venus alignments of possible agricultural importance" (Aveni, 1982a: 22). These provisional indications suggest a pattern that the geography of science would predict for a society of hydraulic farmers.

But were the Maya hydraulic farmers? If they were not, the chain of reasoning would be broken. A visitor to Maya country today finds dense jungle that, if suited for agriculture at all, appears to sustain only a nonintensive swidden (i.e., slash-and-burn) technology. And, indeed, the living descendants of the Maya are swidden farmers in their ancestral lands. Moreover, the density of the jungle conceals, at least from a casual observer on the ground, all traces of any extensive hydraulic infrastructure that may have existed. Until recently, almost every student of Maya society concluded that the Maya had practiced swidden agriculture, despite the theoretical anomaly that food production on the low level of intensity permitted by swiddening should not have been able to sustain a high civilization with monumental architecture, social stratification, and population concentrations in urban centers. In 1946, the Mayanist Sylvanus Griswold Morley wrote:

The modern Maya method of raising maize is the same as it has been for the past three thousand years or more—a simple process of felling the forest, of burning the dried trees and bush, of planting, and of changing the location of the cornfields every few years. This is practically the only system of agriculture practiced in the American wet tropics even today, and indeed is the only method available to a primitive people living in a heavily wooded, rocky, shallow-soiled country like that of the northern Yucatan Peninsula where a plow cannot be used, and where draft animals are not obtainable. (Morley, 1946: 141)

During the period that this swidden thesis prevailed, the inconsistency of a feeble agricultural base supporting a splendid civilization was noted but not stressed. Gordon Willey remarked that "the subsistence base of Maya society was simple. The Maya farmer cleared the jungle with stone tools and fire, planted his maize and bean crop, harvested the crop, and stored it. . . . The extreme simplicity of Maya technological equipment contrasts sharply with the glories of the ceremonial center." (Willey, 1966: 124). And Victor W. von Hagen (1961: 251) puzzled over the failure of the Maya, the "intellectuals of the New World," to develop hydraulic agriculture, and he attributed it to a "Neolithic mental block."

In the face of these authoritative opinions, and in the absence of archaeological or documentary counter-evidence, it called for uncommon confidence in theoretical consistency to assert that Maya agriculture must have reached an intensive, hydraulicized level. Suggestions along these lines were occasionally made, and as early as the 1950s the swidden thesis began to be more seriously questioned (Palerm, 1955; Katz, 1972: 68; see Turner, 1978: 16–17). But most Mayanists clung to it until the 1970s when their complaceny was disturbed by direct evidence of the ancient practice of intensive forms of agriculture. In a familiar pattern of intellectual change, where evidence of hydraulic structures had previously been observed rarely, and generally disregarded, it now was seen almost everywhere and rapidly led to a consensus that "the claim that the Maya represent the world's only swidden-based civilization can now be rejected" (D. Harris, 1978: 301).

In 1974, Billie Lee Turner reported "new evidence of intensive Mayan agriculture, in the form of terraces and raised fields" (Turner, 1974: 118). Two years later, Ray Matheny described the results of aerial photography of the Yucatan Peninsula taken during the wet season with infrared film. The study revealed an elaborate ancient system of waterworks at Edzna that had required "approximately 4 million man-days of labor" and that, he believed, had "figured importantly in some form of intensive agriculture" (Matheny, 1976: 643, 646). Another survey of Yucatan, under the direction of R. E. W. Adams, used a radar mapping system that the Jet Propulsion Laboratory of Pasadena, California, had designed for a study of the surface of Venus. Adams' analysis of the results showed extensive networks of canals the major function of which "was apparently drainage as much as irrigation" (Adams, 1980: 209). The hydraulic characteristics of Maya civilization were further confirmed by the location of major Maya centers on the edges of swamps, and by the fact that the swamps seemed "to have been drained, modified, and intensively cultivated in a large number of zones" (R. E. W. Adams, 1980: 206, map on 210).

One Maya hydroagricultural habitat, with the beguiling name of Pulltrouser Swamp, has recently been closely studied. Ground surveys of the site, which lies in northern Belize, were conducted in 1979, and a full, fine-grained report, along with cautious and provisional interpretations, was published in 1983 (Turner & Harrison, 1983).

Pulltrouser Swamp is a depression in the wet lowlands of the Yucatan Peninsula where, it now appears, the Maya, beginning more than two thousand years ago, created "a complex hydraulic work" consisting of raised fields criss-crossed by canals and designed to provide both drainage and irrigation: "At Pulltrouser the evidence suggests that a drainage-irrigation function was employed and that this may have been true for fields located in similar habitats throughout the area" (Turner & Harrison, 1983: 257).

The hydraulic engineering that produced the system evidently triggered a centralization of authority, for the system could not have been constructed piecemeal without bureaucratic supervision: "The hierarchical canal network would seem difficult to produce without a master plan. Furthermore . . . the system may not have worked properly without total implementation" (Turner & Harrison, 1983: 259). In the super-humid conditions of Maya agriculture in the Yucatan lowlands, large labor forces and their organization and management by a central authority were required not only for building canals, reservoirs, and drainage systems, but also in connection with the technique of raised-field cropping that the Maya employed. Raised fields, which were used by the Maya and the Aztecs to cope with excessive moisture (and are receiving increasing attention from modern farmers today), necessitate high infrastructural start-up costs and considerable maintenance, comparable to the more common riverine irrigation in arid regions.[14] Estimates of the labor required to construct the system at Pulltrouser Swamp range from seven to thirty-three years for a gang of one hundred workers, and "there is no doubt that construction costs were high by absolute or relative measures" (Turner & Harrison, 1983: 260, 267).

Maya studies have confirmed still another characteristic feature of hydraulic societies—urban development: "There is now little question that Maya centres were functionally true cities," possessing "all the necessary urban components, including monumental architecture of various administrative uses, writing systems which record the histories of elite classes and centres, occupational specialists who served the needs of the elite, and large numbers of supporting people" (Adams, 1980: 206). At the peak of Maya culture there were more than forty cities, each of which housed at least twenty thousand people (Paz, 1987: 3).

Urban centers, raised fields, terraces, irrigation and transportation canals, reservoirs, drainage systems, and dams, all requiring substantial inputs of labor—these are the accomplishments of the hydraulic engineering and hydraulic agriculture of pre-Hispanic Maya society that Mayanists have now come to accept. In a report of new research on Maya history, Mayanists Linda Schele and Mary Ellen Miller singled out the development of hydraulic agriculture as the chief factor in the formation of Maya civilization during the Middle Preclassic period (900–300 B.C.): "Perhaps most important for the development of civilization, the Middle Preclassic period saw the beginning of intensive agriculture in raised-field systems and the development of major water management programs at sites like Edzna" (Schele & Miller, 1986: 26). And in a preface to their book, Michael Coe concluded that "it has now been determined that the Maya *did* live in real cities, and that while they carried on milpa or slash-and-burn agriculture, as advocated in the well-known books of Sylvanus Morley and Eric Thompson, there was widespread and highly productive intensive agriculture in the swampy *bajos* that characterize the southern Lowlands" (Coe, 1986: 2).

The most recent research on the Yucatan Peninsula supports the thesis that complex society developed in step with intensive agriculture requiring hydraulic engineering and management. New excavations at Nakbe in northern Guatemala indicate that Maya urban centers and intensive agriculture began to develop around 600 B.C., four hundreds years earlier than had been thought. David Freidel, an authority on Maya archaeology, gave his opinion that "further research will show the Maya were converting the swampland to farming, with raised croplands surrounded by drainage ditches, and that the urban ceremonial centers were manifestations of a central power connected to the management of intensive agricultural systems" (quoted in Wilford, 1989).

The demise of the swidden thesis of Maya agriculture has filled out the picture of a hydraulic society in the Yucatan lowlands, but it has not clinched the argument. The investigators of Pulltrouser Swamp concluded cautiously, "The evidence continues to mount that the Maya were on a par with other early civilizations in terms of agrotechnology. Whether or not this evidence adds strength to theories of sociopolitical development that emphasize agrotechnological controls is the subject of future work" (Turner & Harrison, 1983: 267). At Edzna, however, the elaborate hydraulic system that has been revealed by recent research has permitted less equivocal conclusions. Ray Matheny asserted that:

The hydraulic system of Edzna was apparently a functioning complex, designed to collect, store, and distribute water to a city, rather than the simple sort of system found at Santa Rosa Xtampak. Planning, execution, maintenance, and additions to the Edzna hydraulic system called for much forethought and coordination of peoples. It is easy to conceptualize a central authority on the city level responsible for this type of activity. (Matheny, 1978: 209)

And he concluded that, "if Wittfogel's terminology is acceptable, then a hydraulic society existed in the Maya lowlands at least from the Middle Preclassic period onward" (Matheny, 1978: 207).

The revised understanding of Maya civilization removes a historiographical singularity. The recognition that the Maya economy was substantially supported by a hydraulic infrastructure now lends further credibility to correlations between zones of hydraulically intensified agriculture and the development of a bureaucratic organization of society and its scientific institutions.

SNAKETOWN AND SUN DAGGERS

Maya society presented the most difficult case for archaeologists to analyze among the Sun Kingdoms of Meso- and South America. Until its wetland ecology was decoded it seemed to be unique among high civilizations and seemed to falsify the hydraulic hypothesis. Now, on the contrary, by revealing an Asiatic economy on the Yucatan Peninsula it lends strong support to that interpretation of ancient society and history. Similar conclusions supporting the hydraulic hypothesis had previously been reached for the high civilizations both of central Mexico, where an intensified lacustrine agriculture was practiced by the Aztecs (Palerm, 1955; Coe, 1964; Armillas, 1971), and of the western coastal strip of South America where transverse rivers debouching from the Andean highlands created natural oases, which were artificially irrigated and their fields fertilized by the inhabitants with guano and fish manure (Collier, 1955; Wenke, 1980: 651; Salaman, 1985: 34). Using diverse agricultural technologies, the Aztec of Mexico and the Inca of Peru intensified their agriculture by hydraulic techniques and, like other hydraulic civilizations, institutionalized their astronomy and higher learning to serve the interests of the state and society.

But, the high civilizations of Central and South America do not tell the whole story, for the arid and swampy zones of the New World are not confined to the territories south of the Rio Grande. North of Mexico, in the desertified regions of the American West, there are sites where hydraulic agriculture has been established. Although Amerindian societies north of the Rio Grande never reached the levels of complexity attained in Central America and Peru, they achieved an intermediate development in which both hydraulic agriculture and astronomy reached levels somewhere between what would be expected of Neolithic farmers on the one hand and fully developed high civilizations on the other.

A prehistoric interest in astronomy is reflected in the many astronomical artifacts scattered all over the North American continent—rock art in California depicting stellar and solar patterns, pictographic records of the supernova of 1054 A.D., solstice observatories in New Mexico at Wizard's Roost and Wally's Dome, the "medicine wheels" for solstice observations constructed by the Plains Indians at Big Horn and Moose Mountain, and the "standing stones" along the East Coast (Williamson, 1981a). Most of those artifacts are consistent with the astronomical interests of nomads, food collectors, and Neolithic farmers in other parts of the world; but in the arid districts of the southwestern United States, in New Mexico and Arizona (the name of which confirms its ecological condition), there appear the signs of an incipient hydraulic society along with an astronomy that is more advanced than that practiced by the hunters and Neolithic farmers of North America.

The Pueblo Indians of the American Southwest achieved the highest level of social complexity of any aboriginal Americans north of Mexico.[15] Some of their cultural attainments were derived from their more advanced neighbors to the south, and the degree to which their development was the result of diffusion or independent progress cannot, at present, be accurately estimated. However, at a very early period they were practicing irrigation farming, which they derived from Mexico, and at the peak of their development they were living in multistoried houses.

The agriculture of the southwest Pueblos has been studied most extensively at Snaketown, in Arizona, where the oldest known irrigation system in North America was found. The site is presently occupied by the Pima Indians who regard themselves as the descendents of the original inhabitants, the Hohokam—"the ancient ones."[16] Although it has been known since the 1880s that irrigation agriculture was practiced by ancient populations of the Southwest, it was first extensively demonstrated by excavations at Snaketown in 1934–35 (Gladwin et al., 1938). Since at that time the hydraulic technology of ancient Mexico had not yet been uncovered, the archaeologists at Snaketown drew the conclusion that the system of irrigation farming they had found was an indigenous creation of the Hohokam. Only in the 1960s, when Emil Haury, who had participated in the earlier project, supervised a second program of excavation, did the picture of a derivative hydraulic society in its formative period come fully to light. It then became evident that the Hohokam had migrated from Mexico in full possession of a hydraulic technology and that around 300 B.C., much earlier than had been thought, they constructed an irrigation infrastructure, as soon as they set foot in their new habitat.[17]

In an irony that is now familiar, a desolate, dehydrated site was selected precisely for its hydraulic potential. Native food collectors who inhabited nearby territories had avoided it because it could not sustain an exclusively hunting-and-gathering economy; and with mean annual rainfall of less than 10 inches, neither could it sustain dryland farming. But the original Hohokam settlers, with the practiced eye of irrigation engineers, correctly appraised the agricultural possibilities of the site, which they and their descendents then prosperously inhabited for fifteen hundred years. They saw that the Gila River could be diverted through a canal to water arable fields on two natural terraces. The lower terrace between the river and the upper terrace was relatively easy to irrigate, but it provided little acreage. The upper terrace offered extensive tillable lands, but it required a diversion several miles upstream and a canal along the lip of the terrace.

Accompanying these hydraulic engineering projects, which by Old World or Mesoamerican standards were small, was a comparable degree of political centralization. The construction of diversion dams and canals and the maintenance of the system obviously required some measure of coordinated effort and, insofar as the distribution of water touched the interests of more than one community, some form of regional authority. The relatively modest social constraints required to manage the system and to allocate the water never required institutions as coercive as those that full-fledged Asiatic economies evolved in the Old and New worlds. Nevertheless, the intensified agriculture of the Hohokam resulted in population growth and territorial expansion along with a degree of political centralization commensurate with the level of material culture:

The increased population expanded to occupy a very large territory, and the Gila-Salt region probably attained the highest population density in the Southwest during the prehistoric period. Along with this dynamic growth pattern we see a trend toward increasing complexity in settlement patterns. Large sites possessed a wide range of public facilities such as ball courts and platform mounds, and many settlements were integrated into larger social and economic networks through their participation in irrigation projects of sufficient magnitude to have required a somewhat complex form of political organization. . . . Surely some form of centralized decision-making was necessary for the construction, maintenance, and use of the irrigation systems, ball courts, and platform mounds. (Doyel, 1979: 550)

The archaeological sequence being revealed in the arid Southwest is that of an Asiatic Mode of Production in the process of formation. For reasons that are still obscure, Snaketown was abandoned in the thirteenth century, and by the end of the fifteenth, the Amerindian societies of the entire Southwest were blighted. In the hydraulic regions of the Old World, the Asiatic Mode of Production left a testament of dead cities; in its less intense development in the American Southwest it left dead and ruined villages.[18] By the nineteenth century, Amerindian successor societies of the Southwest were showing signs of rebirth, but they were struck down again, this time through the device of ecological imperialism by hostile and uncompromising invaders.[19] The fragility of an economy based on irrigation agriculture can be seen in the devastation that overcame the Piman successors of the Hohokam at Snaketown when the hydraulic farming system they were reconstructing was destroyed by upstream settlers who peremptorily violated their water rights.

The Asiatic pattern of Amerindian history in the American Southwest prior to the thirteenth century has recently been confirmed by the discovery of an appropriate astronomical development. The intermediate level of agricultural intensification and social complexity attained by the ancient Pueblos was matched by an intermediate level of astronomical interest and accomplishment. In 1977, an unexpected discovery in the area of the Southwest that had been the homeland of prehistoric irrigation farmers showed that they had attained a level of astronomical and astrological interests corresponding to their level of agricultural development.

The Anasazi Indians of the Pueblo culture-complex occupied sites in the Four Corners region of the Southwest where the boundaries of Arizona, Utah, Colorado, and New Mexico intersect. They flourished later than the Hohokam and evidently derived some of their cultural achievements from them, including potterymaking and the cultivation of cotton (Haury, 1976: 355). They also practiced irrigation agriculture, and in several ways displayed traits of an even more developed hydraulic society than the Hohokam. By the early 1970s there had been found evidence of an "agricultural-ceremonial-calendrical system used by the Pueblo Indians of Chaco Canyon [the Anasazi] a thousand years ago" and of their application of celestial observations "primarily [to] the calendar and the foretelling of omens pertaining to the important economic matter of crops" (Ellis, 1975: 59, 84). The new discovery in 1977 would confirm and strengthen this impression of an astronomical system significantly more advanced than those achieved by the surrounding nonhydraulic cultures of North America.

Among the most likely sites for an Anasazi observatory was Fajada Butte, rising 450 feet above the valley floor, in Chaco Canyon, New Mexico. However, until the 1970s, despite careful inspections, no solstice markers or any other signs of an observatory had been found on the butte (Williamson, Fisher, & O'Flynn, 1977: 206–7). Then, in June of 1977, a few days after the summer solstice, a group of amateur archaeologists led by Anna Sofaer discovered a unique Anasazi calendar device high up on the butte that reflects a level of cultural complexity beyond that of food collectors or Neolithic farmers but not yet as high as that of mature Asiatic societies.

Sofaer had come upon two spiral petroglyphs, one large and one small, and she watched as a dagger-shaped beam of sunlight slowly moved downward across the large spiral in step with the sun's westward movement (see Figure 3). The sun dagger was cast by sunlight streaming between sandstone slabs propped upright at the edge of the gallery containing the petroglyphs. At noon on the day of the discovery, the dagger passed slightly to the right of the large spiral's center. Subsequent observations confirmed Sofaer's guess that the device was a unique solar calendar. She and her colleagues returned to the butte a year later, on June 21, the day of the summer solstice, and at noon they observed the dagger as it pierced the heart of the petroglyph. They also saw that at the equinoxes, on March 21 and September 21, the dagger crosses the spiral halfway between its center and an edge, while a second dagger bisects the smaller spiral. At noon on the day of the winter solstice, the two daggers bracket the large spiral, lying tangent to its left and right edges.

Not only is the Anasazi sun dagger a highly accurate seasonal marker, but Sofaer and her associates have presented evidence that the device also records the 18.6-year cyclic extremes of lunar declination (Sofaer, Sinclair, & Doggett, 1982). The unlikely possibility that the sandstone slabs had accidentally fallen into place was eliminated when it was shown that they had been cut from a shelf lying off to the side of the petroglyph gallery (Frazier, 1980: 60).

The sun dagger is the only known calendrical marker that depends on observations at midday, rather than at sunset or sunrise. It attests to an intense interest in astronomy and to an organized effort to conceive and construct the elaborate device, which serves only calendrical and perhaps astrological purposes. Moreover, the period during which the sun dagger was constructed, 950 to 1150 A.D., coincides with a time of thriving economic and political development in Anasazi history. According to Sofaer, "during this phase of extensive trade and ceremonial activity in the canyon, complex systems of roads, communications, irrigation, and ceremonial architecture were developed. The largest structure of the entire Anasazi culture, Pueblo Bonito, . . . a five-story, eight hundred–unit building, was built with its primary elements of design precisely aligned to the rising and setting of the equinox sun" (Sofaer, Zinser, & Sinclair, 1979: 290). The Chaco Canyon road system, consisting of hundreds of miles of roadways, is unmatched anywhere in prehistoric North America, and aerial photography has revealed communities much larger than

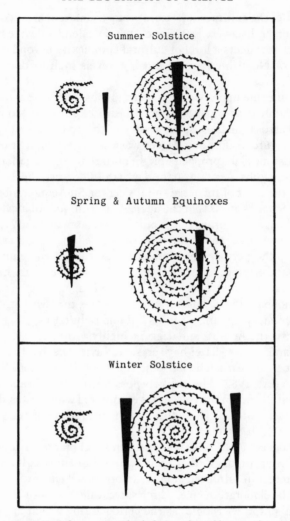

Fig. 3. The Anasazi "sun dagger," in which dagger-shaped beams of sunlight mark the solstices and equinoxes on petroglyphs on Fajada Butte, in New Mexico. At the summer solstice one dagger passes through the center of the large spiral petroglyph. At the equinoxes the other dagger passes through the center of the small spiral. At the winter solstice the daggers bracket the large spiral.

had been suspected. Summarizing these findings, the opinion has been stated that "the canyon may have been a region-wide trade-redistribution center, with full-time administrative specialists running the operation" (Frazier, 1980: 65). Among the latter were astronomers who exploited the high level of Anasazi technical capability for the construction of an observatory. Insofar as monumental building is a measure of social stratification, it is clear that the Anasazi of Chaco Canyon had developed a more complex society than the Hohokam, compared with whom there was more communal construction, more administrative and political centralization, and more institutionalized astronomy.

At the time of its interrupted development, the Asiatic Mode of Production in the Great Southwest seemed to be on its way to rivaling some of the civilizations of the New and Old worlds, recapitulating the early stages of the tripartite development of intensified agriculture, centralized administration, and advanced astronomy. It was undone by an insufficient endowment of surface water, a technology too weak to do more than scratch the earth, and probably by salinization of its soil.

Everywhere across the Old and the New worlds, in all of the earliest state-level societies, some form of hydraulically intensified agriculture formed the basis of the economy. The strong correlation between cultural development and ecological habitat in both the Old and the New worlds strengthens the hydraulic hypothesis measurably. It is an all the more remarkable illustration of the power of ecological imperatives that Amerindian agriculture, despite its isolation from Old World influences and its lack of the plow, the wheel, and draft animals, nonetheless converged with systems found in the Old World, evolving hydraulic societies, complete with their cultural regalia.

But how complete was the isolation of the New World? Before the similarities between the New and Old worlds can be fully counted in support of a geographical interpretation, one possibility needs to be reviewed. The force of the argument would be lessened if the isolation of the New World proved to be more apparent than real—if, that is, there had been regular contact between the Americas and Asia across the Pacific that allowed for the diffusion of Old World techniques and ideas.

Ironically, it is precisely the high level and apparent originality of ancient American civilizations, including their astronomical knowledge, that have evoked suggestions that they were derivative, and have lent plausibility to conjectures of trans-Pacific contacts. Considerable research documenting the similarities between the societies of ancient America and the ancient East, and numerous reports of landfalls by drifting rafts and dismasted junks along the western coasts of North America have strengthened these conjectures. Joseph Needham and his collaborator, Lu Gwei-Djen, recently reviewed the enormous collection of evidence that has accumulated over the past century, and they reached the conclusion that, while "there is a multitude of culture-traits which point to influences from, and contacts with, the Old World," there is "nothing [that] can in any way diminish the profound originality of the Amerindian civilisations, especially in Meso-America and South America" (Needham & Lu, 1985: 2).

If these "trans-Pacific echoes and resonances," as Needham has termed them, prove anything, they confirm an environmental factor in the development of culture, since these allegedly diffused features were absorbed only into the high civilizations of Mesoamerica and Peru and not into any of the nomadic Amerindian societies. Because of the prevailing currents in the Pacific, easterly journeys of disabled vessels would have occurred between the twenty-fifth and fiftieth parallels, and landfalls would, accordingly, have taken place between British Columbia and Northern Mexico (where, in fact, they have been recorded), from whence it is a long way to the homelands of the Aztec, the Maya, and the Inca. Few, if any, of the cultural traits that might have arrived from the high civilizations of Asia across the Pacific appeared in the food-collecting societies that lay between the area of the landfalls and the civilization centers of Meso- and South America. The fact alone that the "echoes and resonances" of Old World hydraulic civilizations were produced and heard primarily in the lands of hydraulic agriculture adds weight to the belief that culture corresponds in significant ways to ecological conditions.

The progressive sequence from the Hohokam, with their relatively modest hydraulic system, to the Anasazi, with their more elaborate infrastructure and more developed astronomy, and to the Maya, with their high civilization and their advanced number system and complex calendrical astronomy, suggests that, as they develop, irrigation agriculture and political centralization interact, each inducing progressive developments in the other. Anthropologists, archaeologists, and prehistorians who have tested the thesis have frequently seen degrees of hydraulicization and corresponding degrees of centralization, rather than extreme forms of either despotic or egalitarian societies.[20] The centralized institutions need not be the pharaonic leviathans that were the primary targets of Wittfogel's analysis and animosity.

The criticism of the hydraulic hypothesis that has been directed against intemperate formulations of the thesis is guilty of attacking a straw man. Such criticism does not undermine the extremely suggestive correlation between intensified agriculture in hydrologically distressed regions and authoritarian institutional patterns. Thus, when Leach questioned Wittfogel's immoderate formulation of the thesis, he may have restrained the tendency of the argument to march too far in advance of the evidence. But extreme renditions of the thesis cannot justify a total rejection of the geographical interpretation. Since the 1950s the hypothesis has been subjected to close scrutiny from many angles, and it has been measured by numerous scholars against evidence drawn from many societies. In a study of Mexican prehistory, Eva Hunt arrived at a moderate revision of the hypothesis that avoids its exaggerated claims and defends its analytical integrity:

To reject irrigation as a highly significant variable . . . is to be equally prejudiced. In a sociocultural system, change in one variable, for example, the change to a more successful agricultural technique, is concatenated with readjustments and changes in other variables in the system. If we take this anthropological truism into account we can propose that irrigation may have played a major role as a *necessary* rather than a *sufficient* cause for the emergence of civilization and large city-states with imperial proclivities in the Mesoamerican highlands. (Hunt, 1972: 246)

When all of its components are brought into focus the ecological argument indicates that in the ancient societies of the Old and New worlds, societies that Marx referred to as Asiatic and others have called hydraulic, science displayed a characteristic structure. It was more highly developed and bureaucratized than in any pre-state society and was fundamentally unlike the unpatronized savants, with their theoretical inventories, of classical Greek science. The extent to which the alignments between geographical conditions and the development of science hold in the Asiatic Mode of Production has been indicated in case studies of Ceylon, China, the lowland Maya, and the Southwest Pueblos. A more exacting test of the geographical approach to the history of science will be a review of the ecology-society-culture formation of ancient Greece—rain-watered, decentralized, and productive of a scientific tradition distinguished not by practical knowledge in the service of the state but rather by the play of ideas.

===

CHAPTER THREE

SCIENCE AND CULTURAL AMBIGUITY

SCIENCE AND RAINFALL

When Herodotus visited Egypt in the fifth century B.C. he described to Egyptian priests the Greek system of rainfall agriculture:

On hearing that the whole land of Greece is watered by rain from heaven, and not, like their own, inundated by rivers, they observed—"Some day the Greeks will be disappointed of their grand hope, and then they will be wretchedly hungry;" which was as much to say, "If God should some day see fit not to grant the Greeks rain, but shall afflict them with a long drought, the Greeks will be swept away by a famine, since they have nothing to rely on but rain from Jove, and have no other resources for water." (Herodotus, 1910, vol. 1: 116–17 [bk. 2, par. 13])

Although for the Nile, too, it was rain that ultimately replenished its waters, the rain fell far away in Ethiopia and central Africa. The priests, living in the valley of the Nile, rarely saw rain, and it must have seemed to them exceedingly foolhardy to hazard the economic basis of a society on such a stinting and fickle element.

Greek farmers did not in fact simply rely on rain to fall in adequate amounts in the right places at the right times; they also grappled with the ecological defects of their habitat by means of irrigation, drainage, and reclamation projects, where these were feasible. But in a country where the topographical relief was sharp and level land was scarce, where fertile plains and plateaus were separated by mountains, and where no perennial stream could provision an expansive flood plain with water and silt, these remedies provided only scattered and local solutions. Unlike the great river valleys and flood plains of the East, the Greek landscape was fragmented, and so too were its social and political structures.

The differences between Egyptian and Greek agriculture were indeed great, and different rainfall patterns were, as Herodotus's hosts observed, significant determinants of the differences. But not the only ones. Egyp-

68

tian agriculture was (until the intervention of industrial civilization in the twentieth century) virtually indestructible.[1] Conversely, Greek agriculture had been seriously damaged before the rise of classical civilization, mainly by the effects of Neolithic deforestation (van Andel & Runnels, 1987: 8). Even in its pristine state mainland Greece had been unsuited to sustaining a large agricultural society. Most of its surface is tilted, making agriculture, and particularly the retention of topsoil, difficult; and much of whatever level land is arable is improperly watered by either rain or runoff from melting snow. In some areas total rainfall is adequate for dryland farming, but its timing is unfortunate, summer drought being common everywhere. Moreover, the eastern, Aegean side of the peninsula, which has always supported the largest and thickest population concentrations, lies in the rain shadow of the Pindus Mountains. In the Mediterranean climatic zone, atmospheric moisture is brought in by westerlies from the Atlantic Ocean and is deposited on the western slopes of elevated land, leaving the eastern slopes and plains relatively dry (Semple, 1931: 462). Among the regions of continental Greece, Attica is the driest, receiving average annual rainfall of sixteen inches, which is barely adequate for cultivating grain (Cary, 1949: 75). Even in the prehistoric period inadequate, ill-timed, or ill-placed rainfall was augmented for farming purposes by redirecting its runoff and concentrating it by means of impounding and irrigation, albeit on a much smaller scale than in the Fertile Crescent.[2]

The damage done by early farmers outweighed any improvements they may have made and bequeathed. By the classical period the Greek agricultural habitat was extensively despoiled, and its fields of arable land considerably reduced as an indirect result of widespread deforestation.[3] By the time of Pericles (fifth century B.C.) Greek forests had been largely denuded, and Attica in particular was in the grip of a timber famine. Athens was importing wood for its vital ship-building industry, and the government had prohibited the export of timber (Carter & Dale, 1974: 100; Hyams, 1976: 92).[4]

The removal of trees from sloping land resulted in the desiccation of the topsoil, which was then either blown away, or washed down into bottomlands, or carried by streams to the deltas and the sea. The arable area shrank, and what was left of it was often so poor and its soil so thin that the limestone base was exposed. Inevitably, the destruction of the soil depressed the standard of living. Where their ancestors had eaten meat, the population of fifth-century Greece was eating grain, the bulk of it imported. It has been estimated that Attica, which was agriculturally one of the most impoverished of the Greek states, produced enough wheat and barley for only one-fourth to one-third of its population (Cary, 1949: 76). A hard reality of early Greek history was a food shortage that drove Greeks off their peninsula and forced them to colonize the eastern Mediterranean. There is evidence that population pressure against the food supply had begun before the fifth century; and, as a result of the pressure, waves of Greek emigrants washed over the Aegean Islands and splashed against the coasts of the Mediterranean and the Black Sea (Rostovtzeff, 1941, vol. 1: 92). From the eighth to the sixth centuries B.C. only Sparta, Thebes, and Attica were exceptions to the pattern of emigration, the first two because of their untypical fertility and Attica because its poverty was so extreme that it was still underpopulated.

The Greek success in turning their predicament into glory and grandeur has been explained in various ways. In accordance with conventional ethnocentric attitudes, the Athenians themselves may have been inclined to attribute their eventual ascendancy (but not the original spoliation of their landscape) to their ethnic qualities. Thucydides, in a departure from this conventional wisdom, suggested that the prosperity of Attica sprouted rather from its very poverty—"the soil was [so] poor and thin" that it was not worth conquering and hence "enjoyed a long freedom from civil strife." Moreover, it became a sanctuary for talented dissidents: "For the leading men of Hellas, when driven out of their own country by war and revolution, sought an asylum at Athens" (Thucydides, 1900: 2–3 [bk. 1, par. 2]). Eventually, to satisfy the basic food requirements of a growing population, Athens turned to colonization, commerce, and conquest.

The steps by which the Athenians accomplished their "miracle" provide some understanding of the social and political organization of Greek society. When Greek farmers exhausted their topsoil they turned to subsoil agriculture, cultivating the vine and the olive tree. All of Greece lies within the region where the vine will grow, and Attica lies just below the northern limit of the olive tree (Braudel, 1975, vol. 1: 232). Frequently, the predominance of olive and grape cultivation and sheep and goat husbandry is an indicator that the topsoil has been wasted to the extent that it cannot produce bread grain. A recent archaeological survey of the Argolid peninsula on the Peloponnese has shown that, contrary to the opinion that the early farmers preferred the domesticated olive and grape, their agricultural system was at first "an economy based predominantly on the cultivation of grains," and the olive and grape were introduced only later (van Andel & Runnels, 1987: 86–92). Although olives and grapes and their derivative products (mainly olive oil and wine) provide most nutrients, they do not constitute a complete diet. Hence Greece became economically dependent on distant lands: "The chief characteristic of the economic life of the Greek city-states, especially those of continental Greece and the Islands, was their dependence on other regions. Few of them were economically self-sufficient, in the sense of producing sufficient food for their population" (Rostovtzeff, 1941, vol. 1: 91).

Attica developed a far-flung commercial network, protected by a large and powerful navy, through which they exchanged olives and grapes and their derivatives for wheat. For Attica and several of the other Greek states the grain trade became a high order of governmental priority: "From the time of Solon [seventh to sixth centuries B.C.] the importation of foreign grain into Attica was a matter of such importance as to require state regulation" (Cary, 1949: 76; also Osborne, 1987: 98). "Their governments became the chief *entrepreneurs* in the grain business," and wheat became the chief contraband of war (Semple, 1931: 354). Indeed, the principal action of the Peloponnesian War was directed at interrupting the Athenian grain supply. Thus, as a secondary consequence of the ruination of their agriculture, several of the Greek states became centrifugally mercantile and militaristic in order to compensate for their agricultural deficits. Ecologically and economically, Greece and the ancient East were binary opposites, for the self-sufficiency of the autarkic economies in the Asiatic Mode of Production was the mirror image of the ever-increasing dependence of the Greek city-states on imports to provide adequate supplies of food, raw materials, and essential manufactured products.

The competition for foreign grain supplies was superimposed on other centrifugal forces operating among the Greek states. In contrast with the Nile Valley, which formed a single ecological entity 10 miles wide and 750 miles long, Greece was topographically compartmentalized: insular Sparta guarded its fertile plain and its wheat granaries on the Peloponnese; the Athenians grew their grapes and olives on their dry and rocky peninsula; Epirus tended its flocks across the Pindus; and everywhere patches of arable land and pasturage fit only for sheep and goats were separated by trackless highlands ("few roads capable of wheeled traffic crossed the mountains from one plain to another" [Andrewes, 1978:9]).[5] Moreover, the topographical fragmentation was reinforced by the ethnic diversity that had been produced by migrations from Crete and across the northern and eastern frontiers of the Greek mainland, with the result that a jigsaw configuration of states maintained their separation even as each achieved a high level of social and political development.

The ancient East was arid, hydraulicized along great rivers, physically and socially centripetal in its ecologically coherent river valleys, agriculturally self-sufficient, and politically centralized. Hellenic Greece, despite its numerous but small-scale irrigation and reclamation projects and its state regulation of the grain trade, became a conglomeration of individual city-states searching everywhere from the eastern Mediterranean to the Black Sea for their means of subsistence. Geographical compartmentalization, military competition, the mainland's relative security provided by the Aegean Sea, agricultural impoverishment so extreme that it might not have repaid the effort of conquest, extensive commerce, a widely scattered diaspora, and ethnic diversity—all of these things conspired to keep the Aegean world fragmented into politically independent states until they finally coalesced under the Macedonian protectorate.

The political leader of a Greek city-state, functioning as a grain magnate who directed the overseas transactions of cadres of independent sailors, traders, ship-architects, engineers, and manufacturers, was less likely to become a despot than a water king, commanding a compact region of intensified agriculture and the obligatory labors of a corvée. If he did gather absolute power, he became no more than the tyrant of a compartmentalized district. The government of a city-state in a region with a limited and eroded agricultural base could never concentrate the enormous wealth that a Pharaoh of a seven thousand-square-mile river valley of hydraulically intensified agriculture could use to patronize a pervasive bureaucracy that bent every social and cultural activity towards the interests of the state.

Greek civilization was not only decentralized, it was also unique. The Asiatic Mode of Production occurred spontaneously several times in the arid and semi-arid regions of the Old and New worlds; but the Greek phenomenon occurred once, and it was abetted by its proximity to the ancient kingdoms. The venerable, culturally developed states of the Near East were within easy reach of Aegean seafarers and merchants, and those famous civilizations served Greece as a museum and a university. Greece was an accident of geography. Nowhere else in the world was there another region so fortuitously placed in space and time. If the basin of the Mediterranean Sea had been in one of its periodic dry states there would still have been the Asiatic Mode of Production—in the Near and Far East and in the New World. But, without the separation afforded by the Mediterranean, there might have been no Greek "miracle." In the eastern Mediterranean, Greek society was separated from, and yet in contact with, the centralized hydraulic kingdoms of Egypt and Mesopotamia. That fine and subtle combination of separation and propinquity enabled the Greeks to begin their remarkable odyssey. If Greek civilization was a miracle, it was one of geography rather than of transcendant cultural qualities or ethnic superiority.

This was the physical setting—those rocky and hilly states—that was to form the environment of a new scientific culture, as different from those of the Oriental kingdoms as the Greek ecology was different from the river valleys, and as the centrifugal decentralized Greek society was from the centripetal centralized monarchies of the East. Without the patronage of any bureaucratic government and without an institutional

structure, Greek science found a unique social niche in which to culti-
vate its extraordinary theoretical innovations. It may not be possible to
reach an understanding of *why* a new scientific culture came into being
in this unique habitat. (If Ionia and Athens had remained as bereft of
science as, say, Carthage and Sparta, would there be any grounds for
surprise?) But once a scientific culture arose in ancient Greece it was
shaped by a society that attached no social value to scientific research or
instruction and provided no support for schools of higher learning.[6] Pericles' description of Athens as the "school of Hellas" is splendidly ironi-
cal.

In classical Greece, science was not part of an economic base as it has
become in modern industrial societies in the form of applied science and
as it had already been, in some measure, in the Oriental kingdoms. Nor,
without institutions, could it form part of a cultural superstructure responsive to socioeconomic requirements. Science floated in a sociological vacuum, even more so than, for example, chess or amateur string
quartets do today. It was the province of individual thinkers, who worked
without the benefits and without the restraints of any institutional affiliations. No scribal schools, no Astronomical Bureau, no libraries, no
universities, no museums, no hospitals, no observatories, no laboratories, no supportive foundations or government agencies—no institutions
of any description intervened between the scientist and society, either to
support his efforts or to direct them towards public service.

In all societies where a scientific culture has taken root—except
Hellenic Greece—scientific research has been enmeshed in a network of
institutions. In the Asiatic Mode of Production scientific institutions
existed as scribal schools and as court and temple research bureaus; and
through them scientific thought, in rational and oracular forms, played a
role in the processes of irrigation agriculture and in the machinations of
governance. In the modern world, science has found applications in industry, medicine, and agriculture and has been sponsored by a high density of institutions. Only in classical Greece did science arise as a significant activity without the support of an institutional superstructure. For
three hundred years, from 600 to 300 B.C., Greek savants pondered alone,
stimulated perhaps by a loose community of colleagues and communication. And they produced a novel intellectual enterprise—the philosophy
of nature. In the Oriental kingdoms science had been one of the tools of
statecraft; in Hellenic Greece it became the work, or rather the pastime,
of *Homo ludens.*

THE SCIENCE UTILITY INDEX

The distinction between *theoria* and *praxis* has become a platitude of our Hellenic heritage. In the ancient East, where knowledge, like labor, was conscripted to serve the government and the economy, no such distinction could have been drawn. Among the Hellenic Greeks it was overdrawn. Ludwig Edelstein singled out as a prominent belief among Greek thinkers the theme that "a life devoted to science is considered a life of happiness. . . . The pleasure sought by all men, yet found by most in the enjoyment of the senses or in the winning of glory, is found by the man of science in knowledge itself" (Edelstein, 1963: 16). While the practical benefits that might result from science were not necessarily repudiated and were even sometimes valued, the outstanding characteristic of Greek scientific thought was its theoretical slant, the pleasure derived by the individual scientist from his intellectual achievements. A corollary of that characteristic of Hellenic science was its political and economic uselessness.

Despite the lack of biographical information about the practitioners of ancient Oriental science, the severely utilitarian quality of their work is well confirmed by all available evidence. By comparison, Greek science is biographically rich, and it should be possible on the basis of records of individual achievement to construct a "science utility index" in order to confirm the impression of a scientific culture that was largely detached from any social or economic objectives. In a curious episode in the historiography of science, a utility index was constructed for modern European science in an attempt to resolve a dispute provoked by a Soviet philosopher of science. A similar method may be applicable to the science of the ancient Greeks.

In 1931, just a few months after the Leningrad conference at which the Asiatic Mode of Production was, as it were, expelled from the Communist Party, a Soviet delegation journeyed to London to attend the Second International Congress of the History of Science and Technology. Among the delegates was a philosopher, Boris Hessen, who presented a paper with the captivating title "The Social and Economic Roots of Newton's 'Principia.'" It became the centerpiece of the conference, and the issues it raised are still being debated. Essentially, Hessen attempted an analysis of Newton's great book in terms of the Marxist categories of economic base and cultural superstructure. "The method of production of material existence," he said, "conditions the social, political and intellectual process of the life of society" (Hessen, 1971: 152). He related the themes of Newton's *Principia*, along with its choice of problems, to the economic pattern of an emergent capitalism in England. By citing numerous correlations between Newton's work and its alleged applications to contemporary engineering and economic problems, he attempted to exhibit "the complete coincidence of the physical thematics of the period, which arose out of the needs of economics and technique, with the main contents of the 'Principia'" (Hessen, 1971: 176). However, to avoid the pitfalls of "vulgar materialism," the influence of the economic base must not be used alone to explain the scientific superstructure. In an exercise of dialectical athleticism Hessen shot the causal arrow in both directions at once: science is affected not only by the economic base, but also by "various superstructures," in which he included "theories," "beliefs," and "dogmatic systems": "The economic position is the foundation. But the development of theories and the individual work of a scientist are affected by various superstructures . . . political, juridical, philosophic theories, religious beliefs and their subsequent development into dogmatic systems" (Hessen, 1971: 177).

Hessen failed to recognize that the cultural superstructure, which purportedly is responsive to changes in the economic base, must be contained in institutions, not alone in the ideas of individual thinkers. That common Marxological confusion enmeshed his interpretation of Newton's thought in grave difficulties.[7] He faced the hopeless project of relating "theories" and "beliefs," which are often the creations of individuals without any thought of economic utility, to English society's social and economic "roots" without any institutional intervention. The result was that despite the suggestiveness and provocative originality of his essay his conclusions appear naive and tendentious. Two generations after Hessen read what proved to be both his seminal paper and his swan song, scholars continue to search in vain for that mare's-nest, a sociological principle that would account for those putative direct connections between the ideas of theoretical scientists and their socioeconomic roots.[8]

Quite apart from the merits of his detailed arguments, the strong emphasis Hessen placed on the close relationship between society and scientific thought was bound to foment controversy. Six years after the Second International Congress, the British social historian G. N. Clark delivered a riposte to Hessen's paper. As a non-Marxist he was indifferent to the organization of the argument around the base and superstructure model, and concerned himself mainly with the claim that, as Hessen phrased it, "the brilliant successes of natural science during the sixteenth and seventeenth centuries were conditioned by the disintegration of the feudal economy, the development of merchant capital, of international maritime relationships, and of heavy (mining) industry" (Hessen, 1971: 155; Clark, 1937: 63–64). He objected to singling out the influence of economics alone (despite Hessen's insistence upon the explanatory importance of the cultural superstructure) and concluded that there were, in fact, six "influences" and "motives" behind scientific research—"five different groups of influences which worked upon science from the outside: those from economic life, from war, from medicine, from the arts, and from religion" and "a sixth, and greatest," the "disinterested love of truth" (Clark, 1937: 86, 89). Through this sixth motive—"the impulse of the mind to exercise itself methodically and without any practical purpose"—Clark effectively divided science into its pure and applied dimensions, and implied that the pure component cannot be accounted for by "influences which worked upon science from the outside." He added that the distinction between the two dimensions might be examined by constructing a science utility index, "by tracing through the thought of the sixteenth and seventeenth centuries the distinction between pure and applied science" (Clark, 1937: 86).

Clark's suggestion for a quantitative index was turned against him by his sharpest critic, the American sociologist Robert K. Merton. Writing in the pages of the Marxist quarterly *Science and Society*, Merton sprang to Hessen's defense. He was critical of Clark's emphasis on judging the work of scientists by their consciously held objectives, the "disinterested love of truth." Instead, he stressed the importance of the "social rôle played by their research" (regardless of their conscious motivations), and he concluded, in support of Hessen, that "it seems justifiable to assert that the range of problems investigated by seventeenth-century English scientists [but not necessarily the substance of their thought] was appreciably influenced by the socio-economic structure of the period" (Merton, 1939: 5, 27).

In addition to the illustrative material that he presented in his paper, Merton buttressed his argument by constructing a quantitative index, a survey of the extent to which research by members of the Royal Society in the 1660s and 1680s was either applied or pure—the former, directly or indirectly, related to "socio-economic demands" while "researches which evidenced no relations of this sort were classified as 'pure science'" (Merton, 1939: 24–26). The utility index that Merton constructed was based on research reports in Thomas Birch's *History of the Royal Society* (1756–1757), and it unavoidably required subjective evaluations of the reports in order to classify them. Each research event was classified as either pure or applied, and the utility index was calculated as the ratio of applied research projects to the total number.[9] He conceded that his method, while "feasible," was "in several manifest respects inadequate"; but it enabled him to approximate the ratios of pure and applied research in England during the Scientific Revolution and, thereby, to test portions of the thesis that Hessen had presented eight years earlier.

Merton employed his science utility index to examine the claim, within the confines of a single society (seventeenth-century England), that the choice of research problems sponsored by the Royal Society was "appreciably" affected by the "socio-economic structure of the period." Applying a similar procedure to science in the ancient world it should be possible to confirm the impression that in Hellenic Greece science was substantially useless.

THE CALIBRATION OF USELESSNESS

Greek science did not originate in Greece. It arose on the (then) fertile Mediterranean coast of present-day Turkey, at first in the city of Miletus and later in several other cities of the region then known as Ionia. The first three scientists for whom even perfunctory biographies can be constructed—Thales, Anaximander, and Anaximenes—were Milesians. Of the first eleven scientists (chronologically) in the *Dictionary of Scientific Biography*, nine were Ionians, five of them from Miletus itself.[10] Twelve of the first eighteen were from this region (including one from nearby Halicarnassus), while the other six were from Italy. And of the first twenty, over a period of 165 years, none was Athenian.

Although the concentration of the founders of Greek science in this region of Asia Minor is indeed remarkable it was no mere quirk of provincial prominence, for in the seventh century B.C. Ionia was the center of Greek civilization while the Greek mainland was the province. Miletus was "decidedly the largest Greek city" (Chandler & Fox, 1974: 79). Like peninsular Greece, Ionia was geographically compartmentalized and gave rise to small political units. However, lying on the western side of the Anatolian highlands it received more rainfall from the Mediterranean westerlies than did Attica, and it was also more generously endowed with fertile land. Although the introduction of olive and grape cultivation by the Milesians at the end of the eighth century B.C. suggests a deterioration of the soil, until the seventh century Miletus and the Ionian island of Chios were self-sufficient in food, and at least in Miletus grain remained a major crop and meat and fish were in greater supply than in Old Greece. Even two centuries later Ionia remained economically superior to the Greek mainland: "Ionia was regarded in the fifth century [B.C.] as a country unusually favored by nature with productive land and herds, with trees, water and a good climate. In contrast to the small Aegean islands and to much of Greece, the Ionian cities were well off for pasture land and tillage. . . . It was a lusher and larger Greece" (Roebuck, 1959: 19).

Ionia also nurtured urbanization earlier than did the mainland (Emlyn-Jones, 1980: 7). And, according to Herodotus, it was a center of technological development, as indicated by the fact that three of the greatest engineering achievements of the Greeks during the early classical period—a temple, a harbor mole, and the famous tunnel of Eupalinus—were constructed on the Ionian island of Samos (Herodotus, 1910, vol. 1:

239–40 [bk. 3, par. 60]). After the Persian conquest of Ionia in the sixth century B.C., some Greek commentators attributed the catastrophe to the soft, effeminate habits into which the Ionians had allegedly fallen as a result of their great prosperity and sumptuous luxury (Roebuck, 1959: 3). The record shows that among the luxuries they enjoyed was one that no other society had ever before indulged itself in—natural philosophy.

The pursuit of knowledge for pleasure, with no apparent use, has always provoked ambivalent reactions. Where Oriental science was largely dedicated to economic or medical applications or to predicting the future—either the weather and the flow of rivers for agricultural purposes or the outcomes of state enterprises for political purposes—the studies and speculations of the early Greek sages were directed mainly at abstract knowledge. This unprecedented activity, engaged in by individuals without any institutional endorsement or support, raised among the classical Greeks questions that have ever since received conflicting answers. Was the knowledge indeed useless? If so, what justification can there have been for diverting an enterprise—systematic study—which in the Asiatic Mode of Production had proven its social utility, into channels of play and recreation? In the modern world the cultural ambiguity engendered by pure science, indifferent to world betterment, derives from a long tradition. In the Aegean world it was something new, since previously no Oriental potentate would have sponsored pure science, and unsponsored research was impossible prior to the Greeks, or in any event has left no trace.

While the devotees of the new natural philosophy savored its delights, there were others who took a less appreciative view of an avocation so provocatively useless. Although Hellenic scientists received no public support, their utterly impractical and seemingly meaningless private investigations were almost calculated to excite animosity and ridicule. In *The Clouds*, Aristophanes' most pungently antiscientific comedy, he lampooned at once both the new science and the new bourgeoisie in a few deft lines by opposing a disciple of Socrates and pure science against a character (Strepsiades) whose main interest was fending off the litigious assaults of his creditors:

Disciple: Chaerephon of the deme of Sphettia asked [Socrates] whether he thought a gnat buzzed through its proboscis or through its anus.
Strepsiades: And what did he say about the gnat?
Disciple: He said that the gut of the gnat was narrow, and that, in passing through this tiny passage, the air is driven with force towards the breech; then after this slender channel, it encountered the rump, which was distended like a trumpet, and there it resounded sonorously.
Strepsiades: So the arse of a gnat is a trumpet. Oh! What a splendid arsevation!

Thrice happy Socrates! It would not be difficult to succeed in a law-suit, knowing so much about a gnat's guts! (Aristophanes, 1938: 546 [The Clouds, ll. 150–68])

The cultural uncertainty produced in the ancient world by the advent of abstract secular knowledge is further displayed in the divergent and inconsistent traditions that sprang up around the careers of individual savants and in the contradictory interpretations that modern scholars have placed on them. Thales, whom Aristotle designated the founder of Greek natural philosophy, was variously described by Herodotus, Aristotle, and Plato as an eastern-style prognosticator in the service of the state, an olive oil tycoon, and an unworldly dreamer (Longrigg, 1976: 295–96). Aristotle's report is particularly revealing of the ambiguity with regard to the utility, or the lack of it, of the speculations of the early natural philosophers:

There is the anecdote of Thales the Milesian and his financial device, which . . . is attributed to him on account of his reputation for wisdom. He was reproached for his poverty, which was supposed to show that philosophy was of no use. According to the story, he knew by his skill in the stars while it was yet winter that there would be a great harvest of olives in the coming year; so, having a little money, he gave deposits for the use of all the olive-presses in Chios and Miletus, which he hired at a low price because no one bid against him. When the harvest-time came, and many were wanted all at once and of a sudden, he let them out at any rate which he pleased, and made a quantity of money. (Aristotle, 1941: 1142 [Politics, ll. 1259a5–a20])

Aristotle added, "Thus he showed the world that philosophers can easily be rich if they like, but that their ambition is of another sort."

Precisely the same ambivalence is evident in modern interpretations of Thales' career. The view that his "interest in astronomy was, like his geometry, motivated by practical needs" (Huxley, 1966: 96) is contested by the judgment that "these Greeks [referring specifically to Thales], practical men though they were, had a passion for asking useless questions" (Kitto, 1957: 178). In a recent study of Ionian culture, the classicist C. J. Emlyn-Jones inquired into why Thales would have made the eclipse prediction that Herodotus attributed to him. Emlyn-Jones concluded that Thales' motivation in these matters was in direct opposition to what would have been the case for the anonymous savants in the Oriental kingdoms, where socially programmed institutions would have focused such speculations on affairs of state:

In the Babylonian empire such events [eclipses] were of profound religious and hence political significance and their occurrences were assiduously observed and recorded by priests for whom these happenings were intimately connected with their guidance of the state. The Greek "polis," on the other hand, lacking a ruling priestly caste, a developed astrology and a history of mathematical and astronomical expertise, did not supply the motivation for such religious observations. . . . It is reasonable to conclude that Thales' motivation was probably sheer curiosity about the phenomenon itself. (Emlyn-Jones, 1980: 98)

An exactly opposite conclusion—that Thales' renown was based on his political abilities—was reached by D. R. Dicks after a close scrutiny of the few authentic bits of evidence bearing on Thales' career: "What is the general impression we obtain . . . ? Surely that he had a reputation chiefly as a *practical* man of affairs . . . and the ability to put to practical use whatever knowledge he possessed" (Dicks, 1959: 295).

These conflicting interpretations of Thales' career reflect the uncertainties and confusion that were provoked by the first contacts between Eastern traditions of service to the state and natural philosophers operating as individuals without any bureaucratic or institutional constraints. A similar cultural reaction can be seen in traditional accounts of the work of some of Thales' Ionian successors, in which records of theoretical and abstract teachings are confronted by reports that reflect utilitarian yearnings. Anaxagoras was said to have predicted the fall of a meteorite at Aegospatami and Anaximander the occurrence of an earthquake (Longrigg, 1976: 296, 298). Shades of Oriental divination. If only the new science could be made as useful as the old.

But it could not. For there was no institutional structure that could have produced a social agenda for science prescribing socially relevant missions for research. The natural philosophers of Ionia were on their own. Ancient Greek civilization prior to the Macedonian ascendancy late in fourth century B.C. provided the only setting in the history and geography of science in which research was so completely free of institutional authority. The savants made the most of it. They conducted their studies as though they were indeed playing chess, with as little commitment to social betterment or political service. The records of their abstract speculations have been preserved by essentially reliable, albeit fragmentary, reports. According to those fragments, Thales proposed a cosmos composed ultimately of a watery matter; Anaximander theorized that all comes from the infinite and returns to it; Anaximenes declared that Anaximander's infinite was, in fact, air; Anaxagoras insisted that matter, while infinitely divisible, retains its characteristics to the very end; Pythagoras calculated that all things are understandable as numbers—except the square root of 2; Heraclitus sensed that the world was in tension between opposites that were, however, the same; Leucippus thought that the world must be composed of atoms; and Zeno proved that Achilles was not as fleet of foot as a tortoise. What a gallery of delightful whims, of no worldly use whatever beyond the scope of natural philosophy. A grand master would understand.

If the survey of early Greek scientific biography is enlarged to include all scientists prior to the founding of the first Greek scientific institution, the Alexandrian Museum (275 B.C.),[11] this assessment of uselessness can be confirmed. The *DSB* contains fifty-one entries of scientists (all Greek) who flourished before Euclid composed his *Elements* at the museum; and by following the criteria Merton used in designing his utility index for seventeenth-century England, the main thrust of their work may be separated into the categories of pure and applied research. Obviously, a subjective judgment is again required in reviewing and classifying these biographies. But, to reduce the possibility of overestimating uselessness, medical and anatomical writings, geography, and political doctrine were considered applied research even when there was a slant towards theoretical interests. With every precaution taken, the index of utility for Hellenic science is less than 20 percent (19.6% — ten out of the fifty-one individuals). These scientists, their principal field of study, and the classification of their research are listed in Table 2.

Greek savants undeniably conducted useful studies of mapping, medicine, mensuration, and navigation; but, compared with the scientific cultures of Egypt and Mesopotamia, the component of abstract natural philosophy in Hellenic science was substantially larger. Where the utility index of Greek science is less than 20 percent, in the societies of the ancient Orient the index would approach 100 percent, since by all accounts the institutional authorities showed little interest in sponsoring any higher learning that had no clear potential to produce practical benefits. In Hellenic Greece the internal, cognitive content of science, the specific theories, were for the most part neither productively nor predictively useful; they were carried in no institutional superstructure and had no roots in any economic base. The decentralized structure of Greek society limited the possibility of holding Greek science to any state-dictated agenda. Indeed, the abstract purity of Hellenic science reflects a general and characteristic tendency of the scientific enterprise itself: whenever theoretical scientists slip their institutional leash they are likely to be found in the vicinity of pure, not applied, research. The result in the case of classical science was a dazzling tradition of originality unrelated to any social objectives.

Nonetheless, attempts have been made to account for the internal "positive content" of early Greek science in terms of the socioeconomic features of Greek society. Farrington, in his original, influential, and generally informative survey of Greek science, was one of those who adopted an interpretation of economic base and cultural superstructure

TABLE 2. UTILITY OF RESEARCH BY HELLENIC SCIENTISTS,
585–300 B.C.

Flourished (Years B.C.)	Name	Main Field	P (pure)/ A (applied)
585	Thales	Natural Philosophy	P
570	Anaximander	Astronomy	P
545	Anaximenes of Miletus	Philosophy	P
535	Xenophanes	Political Philosophy	A
520	Pythagoras	Mathematics	P
500	Heraclitus of Ephesus	Natural Philosophy	P
500	Hecataeus of Miletus	Geography	A
495	Alcmaeon of Crotona	Medicine	A
485	Leucippus	Natural Philosophy	P
475	Parmenides of Elea	Natural Philosophy	P
460	Anaxagoras	Natural Philosophy	P
452	Empedocles	Natural Philosophy	P
450	Hicetus of Syracuse	Astronomy	P
450	Zeno of Elea	Philosophy	P
440	Philoaus of Crotona	Philosophy	P
432	Oenopides of Chios	Astronomy	P
430	Herodotus of Halicarnassus	History	P
425	Hippocrates of Chios	Mathematics	P
425	Theodorus of Cyrene	Mathematics	P
420	Democritus	Physics	P
420	Euctemon	Astronomy	A
420	Hippocrates of Cos	Medicine	A
420	Meton	Astronomy	A
420	Antiphon	Mathematics	P
400	Hippias of Elis	Philosophy	P
390	Leo	Mathematics	P
387	Plato	Philosophy	P
380	Leodamos of Thasos	Mathematics	P
377	Theaetetus	Mathematics	P
375	Archytas of Tarentum	Philosophy	P
375	Bryson of Heraclea	Mathematics	P
370	Theudius of Magnesia	Mathematics	P
368	Speusippus	Philosophy	P
360	Eudoxus of Cnidus	Astronomy	P
355	Xenocrates	Philosophy	P
350	Dinostratus	Mathematics	P
350	Heraclides Ponticus	Astronomy	P
344	Aristotle	Natural Philosophy	P
340	Aristaeus	Mathematics	P

(continued)

TABLE 2.
(*Continued*)

Flourished (Years B.C.)	Name	Main Field	P (pure)/ A (applied)
340	Menaechmus	Mathematics	P
330	Aristoxenus	Harmonic Theory	P
330	Callippus	Mathematics	P
330	Phytheas of Massalia	Geography	A
330	Theophrastus	Botany	P
325	Eudemus of Rhodes	Philosophy	P
325	Thymaridas	Mathematics	P
310	Dicaearchus of Messina	Geography	A
301	Epicurus	Natural Philosophy	P
300	Autolycus of Pitane	Astronomy	P
300	Diocles of Carystus	Medicine	A
300	Praxagoras of Cos	Anatomy	A

Source: Based on data in *DSB.*

in which ideas, rather than institutions, are directly related to economic and technical conditions. Referring specifically to the Milesians, he stated:

> They were observers of nature whose eyes had been quickened, whose attention directed, and whose selection of phenomena to be observed had been conditioned, by familiarity with a certain range of techniques. The novelty of their modes of thought is only negatively explained by their rejection of mystical or supernatural intervention. It is its positive content that is decisive. Its positive content is drawn from the techniques of the age. (Farrington, 1980: 41)

As an illustration of a specific idea that was drawn from the techniques of the age, Farrington cited Heraclitus's opinion that all change results from tension between opposites (a metaphysical conjecture that is sometimes claimed to define a research strategy):

> It is one of the richest and most helpful ideas of the old philosophers, not a whit reduced in significance when we remember that it, too, had its origin in the techniques of the time. The doctrine of *Opposite Tension* which Heraclitus applied to the interpretation of nature was derived, as his own words inform us, from his observation of the state of the string in the bow and the lyre. (Farrington, 1980: 40)

Techniques of the time? The bow was first bent in the Old Stone Age and the lyre was plucked in Sumer. Although archery was known, it was little used by Greeks in Heraclitus's time, and was scorned by the Greek warrior (Burn, 1968: 9). Since these instruments had been available for thousands of years, why was the principle of "opposite tension" discovered only around 500 B.C.? That Heraclitus may have drawn inspiration from these ancient devices says something about his mental habits, but it says little about the sociology of Greek science.

Farrington also believed that the first natural philosophers in Ionia were high-spirited entrepreneurs whose vigorous life style somehow contributed to their scientific inventiveness. "Miletus . . . was the most go-ahead town in the Greek world," and "the first philosophers were the active type of man . . . [whose] modes of thought . . . [were] derived from their active interest in practical affairs" (Farrington, 1980: 35). He cited the tradition that Thales may have dabbled in olive oil futures, but he failed to mention either that Anaxagoras "neglected his inheritance to devote himself to natural philosophy" (Longrigg, 1970: 149) or the anecdote recorded by Plato that Thales fell into a well while absent-mindedly stargazing. Since little is known with certainty about the personal lives of these scientists, Farrington's generalization is wholly unwarranted. The most socially significant feature of the traditional accounts of the lives of these early Greek scientists is their ambiguity, reflecting the cultural confusion that their abstract research provoked. In any event, the assumption that businessmen develop a taste for abstract thought through their money-getting is unconfirmed by entrepreneurial history, scientific biography, or common knowledge.[12]

The Ionian social context, with its decentralized economy, had not developed scientific institutions and, hence, left its natural philosophers to speculate about nature in complete bureaucratic isolation, outside of the organization of society. It was not that the Ionian natural philosophers were entrepreneurs, it was that they were individualists. What they dreamed up in their ivory towers cannot be clarified through socioeconomic studies. The thriving and decentralized economies of Ionia may help account for the establishment of a culture of leisured intellectuals who devoted themselves to original thought, but not for the content of the thought itself. Socioeconomic analysis can perhaps indicate that the formation of a coterie of intellectuals was not inconsistent with prosperous commercial cities, sustained by decentralized agricultural bases, and in close contact with the culturally advanced civilizations of the East. Given the rich astronomical tradition of Eastern learning, it should come as no surprise that some early Greek thinkers, familiar with these traditions and in possession of their extensive accumulations of astronomical data, focused their speculations on the nature of the cosmos. Their speculation was for the most part purely abstract and unrelated to either calendrical reckoning or oracular revelation. Had scientific institutions been founded, they would have selected and sponsored specific avenues of research that might have been expected to lead to public benefits. But Hellenic Greece was barren of such institutions.

During the Hellenic enlightenment, which lasted 300 years, no more than perhaps 20 percent of Greek scientists were Athenians or worked in Athens. Their total number probably did not exceed twenty; but among those few was Aristotle. His work was the culmination of Hellenic science. In large measure it set the themes of research across several fields over the next millennium. The extensive writings that we commonly regard as his were compiled to some extent during his lifetime, and perhaps to a greater extent by disciples during the first two centuries after his death. Like all Hellenic scientists Aristotle's research was undirected by any state authority and he had no institutional affiliation. Even the private school generally referred to as his *Lyceum* was probably founded after his death (Owen, 1970: 250). The substance of his studies reflected their sociological status. They were utterly abstract, of no possible use in engineering, medicine, or statecraft. Although Aristotle recognized the distinction between pure and applied research, "speculative philosophers" and "[medical] practitioners," he confined his research to his private interests in the philosophy of nature. Even when he wrote on anatomy and biology, fields that might have lent themselves to useful studies applicable to the treatment of illness, he focused his objectives on the place of living beings in a rational cosmology. Similarly, his studies of the theory of motion, which remained influential until the seventeenth century, were part of a program of purely theoretical research and were of no practical use in technical or economic applications.

When the traditions of Greek natural philosophy and Oriental state-supported institutions were hybridized during the Hellenistic era in the territories of Greek imperialism, and later in Byzantium, Persia, and Islam, it was Aristotle's writings that often represented the Hellenic influence. That merger of the two traditions in the strongholds of the East forms the next chapter in the geography of science.

CHAPTER FOUR

FROM PHARAOHS TO CALIPHS

TO LORDS

THE SYNTHESIS OF EAST AND WEST

Lacking an institutional link between science and society, Greek science in the Hellenic era was left in a condition of social anarchy. That state of affairs was to last only as long as the Greek poleis remained disunited. When the Macedonians asserted their dominance over the Greek states, and specifically, when Alexander the Great (356–323) conquered the eastern kingdoms, the situation changed. In the wake of Alexander's conquests, a merger of the Greek and Oriental scientific traditions followed quickly. After Alexander's death, his satrap in Egypt, Ptolemaios Soter, founded not only a dynasty of Greek pharaohs but also the Alexandrian Museum and Library, reminiscent of the state-supported institutions of the Oriental kingdoms. Through this museum and library at Alexandria and through similar libraries elsewhere in the Macedonian conquest, Greek science was to be orientalized and enter its golden age.

The intellectual history of Hellenistic science, beginning with Euclid, has been frequently studied. The environmental history of its origins is less well known.

Even before the Egyptian priests warned Herodotus that the Greeks would starve if the gods withheld rain, Aeschylus thanked the gods for supplying it: "Holy sky passionately longs to penetrate the earth, and desire takes hold of the earth to achieve this union. Rain, from her bedfellow, sky, falls and impregnates the earth, and she brings forth for mortals pasturage for flocks and Demeter's livelihood" (quoted in Burke, 1987: 7). Nonetheless, the separate regions of Greece supplemented the gift of the gods by irrigating, draining, and reclaiming land in a constant effort to feed their populations and minimize unfavorable trade balances resulting from the importation of food.

The antiquity of the scattered and, compared with those of the East, small-scale hydraulic projects that were undertaken in Greece is attested to by legend, literature, and surviving masonry artifacts. The second labor of Hercules was the slaying of the nine-headed water serpent, Hydra, which grew two heads in place of each that was severed. To destroy the ninth immortal head, Hercules buried it under a huge rock. The Hydra evidently symbolized a stream under high pressure that was difficult to stanch; whenever it was impounded it would spring to life again in another channel, and it was finally controlled only by a massive dam. In several of his labors Hercules is depicted struggling with water control projects of one sort or another (Semple, 1931: 444). Greek water myths, like their counterparts in other ancient agricultural societies, were apparently embroidered accounts of severe ecological problems along with the heroic responses they evoked. Ellen Churchill Semple defended their historical validity:

With deep national insight, the Greeks embodied in their mythology the story of Perseus and his destruction of the sea monster who ravaged the coast, and Hercules' killing of the many-headed serpent who issued from the Lernean Marshes to lay waste the country of Argos. Even so early a writer as Strabo [fl. 23 B.C.] states that yet earlier authorities interpreted Hercules' victory over the river god of the Achelous as the embankment of that stream and the draining of its inundated delta tract by the national benefactor. (Semple, 1911: 327–28)

In the writings of ancient authors there are also less mythological references to riparian rights and to hydraulic projects designed to concentrate and regulate meager hydrogeological resources. Moreover, the survival of some of the actual structures provides definitive evidence that such efforts were made. In Thebes, an atypically fertile region, "several large tanks of ancient masonry with inscribed tablets impound the water of the famous Dircean springs and distribute it to the surrounding gardens. They stand there, still performing their age-long service, mute intimations of the immortality of geographic factors in history" (Semple, 1931: 447). These projects were the work of local water kings; and in conjunction with other public works they promoted regional centralization, albeit on a smaller scale than in the East.

In the fourth century B.C. the most successful water king was Philip II, and Macedonia was the first Greek state to coalesce into a nation. Macedonia fused into a nation in stages. Physically, the province consisted of interior highlands, transected by river valleys, and a coastal plain. As long as the upland valleys remained isolated under their own rulers they were jointly subordinate to whatever authority controlled the plain and, thereby, access to the Aegean. At the end of the fifth century the valleys were connected by roads, and the highland interior came under the governance of a single king, Archelaus. Over the next half-century, consolidation proceeded through strife and disorder until in 359 Philip II, at the age of twenty-three, ascended the throne and completed the process of unification (Botsford & Robinson, 1948: 287–89).

Philip also began the process of expansion that, within another half-century, was to unify most of Greece and create an empire stretching from Cyrenaica to the Indus Valley. Macedonia was then, and remains today, a fertile region if maintained under conditions of irrigation. It also possessed abundant wet meadows along with highland summer pasturage and, consequently, was able to field a large cavalry as part of its armed force (Semple, 1931: 445, 318). In 356 Philip conquered a gold-rich region of neighboring Thrace and undertook the drainage of the swamps in that district. With the substantial gold and silver of Macedonia and Thrace, a sound agricultural base, a large population, the possibility of easily conscripting soldiers in depression-ridden fourth-century peninsular Greece, and a formidable cavalry, Philip—and after his violent death, his son, Alexander—forged the most extensive empire in ancient history (Cary, 1949: 302; Rostovtzeff, 1941, vol. 1: 94–97).

Egypt fell easily to Macedonian arms in 332; and Asia Minor, Mesopotamia, Afghanistan, and western India were conquered soon after. Then, upon the early death of Alexander, in 323, the empire broke up into three major parts—mainland Greece, western Asia, and Egypt—the last garrisoned under the rule of Ptolemaios, who crowned himself king in 305.[1]

Although it was the Greek fashion to build cities as regional centers, Egypt did not lend itself to regionalization the way the geographically fragmented Aegean had, and Ptolemaios therefore built only two cities— one, Ptolemais, in upper Egypt (that is, above the delta) and the other, Alexandria, in the western delta. The main center of Greek life in Egypt was established in Alexandria, where a harbor was built (actually two harbors formed by a single jetty), and where Ptolemaios founded the great museum and library.

Within a century Alexandria became "the greatest city of the known world" (Tarn, 1975: 185).[2] Besides being the administrative hub of an empire, a bustling port, and the preeminent intellectual center in the ancient world, it was also a cultural estuary where East and West mixed in a sparkling interplay of cosmopolitan influences. George Sarton (in a somewhat dated comparison) stated:

Come to think of it, Alexandria must have been a city comparable to New York, the two ruling elements being in the first place Greek and Jewish and, in the second, British (or Irish) and Jewish. And just as New York is a dazzling symbol of the New World, so was Alexandria of the Hellenistic culture. (Sarton, 1959: 21)

The affiliation between the development of higher learning and Macedonian politics had preceded the founding of the Alexandrian Museum;

it originated even before the time of Philip. Aristotle's father served as personal physician to Alexander's grandfather, and Aristotle himself became Alexander's tutor. In Egypt, however, most things were different from the way they were in Greece. In Athens, Alexander's relationship with Aristotle was on a personal and informal basis; in Alexandria, under Oriental influences, patronage of learning was institutionalized by the state. The Alexandrian Museum was entirely unlike anything that had preceded it in Greek history (and it was also unlike the collections of curiosities that the term later came to signify in the European Renaissance). It was, instead, a temple dedicated to the nine Muses, where the members were, in a combination of Greek and Oriental traditions, free to conduct their own research yet fully supported as employees of the state. Although no detailed account of the museum has been preserved, a concise description was recorded by the Roman-era geographer, Strabo:

The Museum is also a part of the royal palaces; it has a public walk, an Exedra with seats, and a large house, in which is the common mess-hall of the men of learning who share the Museum. This group of men not only hold property in common, but also have a priest in charge of the Museum, who formerly was appointed by the kings, but is now appointed by Caesar. (Strabo, 1959: 35 [bk. 17, pt. 1, par. 6])

That the members were supported by the state is further confirmed by the fact that in Ptolemaic Egypt the king appointed an attendant who was "in charge of supplies to the tax-free men who are fed in the Museum" (Rostovtzeff, 1941, vol. 2: 1084). Eratosthenes, who was a distinguished geographer and mathematician, became head of the library, and Euclid taught in Alexandria and may have been affiliated with the museum. Indeed, the institutional support and encouragement that science received in Alexandria evidently contributed to an upsurge in scientific activity through much of the Hellenistic world. (See Figure 4.)

The institutional patronage of science and learning that was centered in Alexandria was no Ptolemaic whim unrelated to the cultural geography of the ancient Middle East. Although the Alexandrian Museum was a unique institution, elsewhere in the Macedonian conquest where Greek culture was planted on Oriental soil the state supported learning either through the establishment of libraries or thorough some other institutional device. The library at Pergamum, founded by a Macedonian dynast, Attalus, upon his inheritance of Persia, was second only to the Alexandrian Library. Antioch, in ancient Syria, under the Macedonian Seleucids, was also hospitable to science and scholarship (Rostovtzeff, 1941, vol. 2: 1084–85).[3] Greek culture alone was insufficient to produce these developments, for in mainland Greece the Macedonian hegemony did not result in any comparable institutional patronage of learning. Only where the Greek and Oriental traditions mixed, only where the

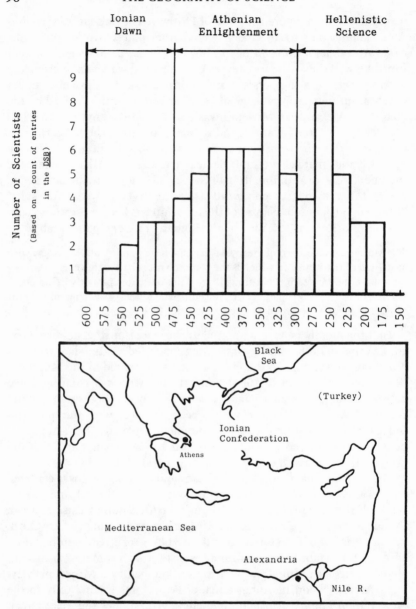

Fig. 4. The geography of Greek science: *Ionian Dawn* (600–475 B.C.): Individual scholars unsupported by any institutional structures in the Ionian Confederation; the earliest scientific biographies. *Athenian Enlightenment* (475–300 B.C.): Science was centered in Athens and dispersed over the central and eastern Mediterranean. *Hellenistic
Science*: With the founding of the Alexandrian Museum in 275 B.C. Greek science was
orientalized in the institutions of the Hellenistic kingdoms.

individualistic scholar was favored by state-supported institutions, did a cultural synergism occur. Although mainland Greece was intermittently prosperous it was never rich. But in Egypt, with its hydraulically intensified agriculture and the opportunity to tax a whole nation instead of merely a city and its hinterland, the central authority was able to gather enormous surpluses and devote some of it to the patronage of science and learning.[4] That the Hellenistic florescence of science occurred in Egypt and the lands of the ancient centralized monarchies and not in the city-states of mainland Greece adds weight to a geographical interpretation of ancient science. Moreover, it was only the scientific culture of Greece that was transplanted to Oriental soil; its decentralized economy and political system were not possible in the hydraulic districts of the East. It is indeed a testament to the influence of material conditions on social life that in the Hellenistic lands where Greek culture took root command economies continued to hold sway. Ptolomaic Egypt produced a dynasty of Greek pharaohs, but never a Pericles.

Then as now, the price of learning was high. And the Hellenic episode had shown that science could not be counted on spontaneously to satisfy either the public's or the state's interests in acquiring useful knowledge. Science needed to be controlled, patronized, and cajoled if it was to do so. It might, therefore, be thought that the Macedonian dynasts had such purposes in mind when they sponsored higher learning. There is, however, little evidence bearing on the specific motives and objectives of the wealthy Hellenistic kings who founded and patronized the state libraries and the Alexandrian Museum. Whatever they may have hoped for in terms of useful knowledge, what the savants they employed actually produced was little more applicable to the solution of practical problems than what the Hellenic philosophers of nature had achieved without elaborate and costly institutional patronage. This first experiment in combining Greek and Oriental scientific traditions resulted in a high degree of social ambiguity.

Although Hellenistic scientists thus preserved a large measure of traditional Hellenic freedom of inquiry, it was probably inevitable that some degree of institutional pressure would be felt by the scholars who were so generously supported by the state through its endowment of libraries and the museum. The utility index of scientific research can be estimated to have risen to 32 percent during the two centuries following Euclid (see Table 3). At that level of utility Hellenistic science would still have been oriented towards pure research, with, however, a substantial and perhaps increasing commitment to the acquisition of useful knowledge. George Sarton may have been too insensitive to the power of institutional pressure when he concluded that Hellenistic research was thoroughly discipline-oriented with no attempt to find applications:

The fellows of the Museum were permitted to undertake and to continue their investigations in complete freedom. As far as can be known, collective research was now organized for the first time, and it was organized without political or religious directives, without purpose other than the search for the truth. (Sarton, 1959: 34)

Hellenic traditions implanted in an Oriental environment can easily be misinterpreted. Among the first fellows recruited from Athens to teach in Alexandria was Euclid. Although it is commonly believed that

TABLE 3. UTILITY OF RESEARCH BY HELLENISTIC SCIENTISTS, 295–95 B.C.

Flourished (Years B.C.)	Name	Main Field	P (pure)/ A (applied)
295	Euclid	Mathematics	P
295	Zeno of Citium	Philosophy	P
290	Herophilus	Anatomy	A
287	Strato of Lampsacus	Natural Philosophy	P
270	Aratus of Soli	Astronomy	P
270	Aristarchus of Samos	Mathematics	P
270	Aristyllus	Astronomy	P
270	Ctesebius	Mechanics	A
264	Erasistratus	Anatomy	A
250	Perseus	Mathematics	P
250	Philinus of Cos	Medicine	A
250	Nicomedes	Mathematics	P
247	Archimedes	Mathematics	P
245	Conon of Samos	Mathematics	P
240	Philo of Byzantium	Mechanics	A
236	Eratosthenes	Geography	A
225	Dositheus	Mathematics	P
210	Apollonius of Perga	Mathematics	P
200	Bolos of Mendes	Biology	A
200	Dionysodorus	Mathematics	P
190	Diocles	Mathematics	A
190	Zenodorus	Mathematics	P
175	Hypsicles of Alexandria	Mathematics	P
127	Hipparchus	Astronomy	P
125	Theodosius of Bithynia	Mathematics	P
110	Zeno of Sidon	Philosophy	P
100	Petosiris (Pseudo)	Astrology	A
95	Posidonius	Philosophy	P

Source: Based on data in DSB.

Euclidean geometry is of great applicability in engineering, it is in fact of only indirect value. The *statements* of Euclid's theorems may be of use to a surveyor or builder, but their *proofs* are not. Even today, it is only the rare engineer who can (or has any occasion to) prove Pythagoras's theorem; and the similar-triangle theorem, the principle of which is not only frequently used but is also transparently simple and almost self-evident, is in fact exceedingly difficult to prove and requires the higher analysis and the method of limits. (In high school geometry courses it is sometimes introduced as a lemma, but it is never proved.) In the ancient world the Pythagorean principle had been known in some sense to the Babylonians (they recorded lists of Pythagorean triplets), and to the extent that it was useful the principle might have been applied in building construction. Euclid's contribution was to supply the proof, a splendid accomplishment for which, however, Hellenistic engineers owed him nothing.

Such was the cultural ambiguity of Hellenistic science. Euclid proved five hundred theorems; Archimedes determined the areas and centroids of parabolic segments; Appolonius discovered the conic sections (which were first applied by Kepler in the seventeenth century to account for planetary orbits); Eratosthenes calculated the circumference of the Earth; and Aristarchus postulated that the Earth revolves around the Sun. If Ptolemaios and his fellow monarchs and successors expected useful knowledge in return for any investment they may have made in those careers, they should have been sorely disappointed by results such as these.

No records remain that can definitively clarify the policies and objectives of the royal patrons who sponsored libraries and research activities in the Hellenistic kingdoms. There seems to have been a shift of limited magnitude towards fields of study that may have been expected to yield useful knowledge—mechanics, hydrostatics, geography, medicine, anatomy, mechanical contrivances, lexicography, and pneumatics—and away from the utterly abstract natural philosophy of the Ionian and Athenian enlightenments. This shift in the choice of research problems may reflect direct or subtle institutional pressures. We cannot know; but we know for sure that, even allowing for a moderate increase of emphasis on utilitarian investigations, the results of the research were generally inapplicable to the problems of economics or governance.

Some historians have drawn the conclusion from slender threads of evidence that Ptolemaios and his successors hoped and expected that their substantial investment in the museum and the library would produce practical benefits in the fields of technology and government. Farrington stated that: "The Ptolemies, in charge of Egypt, would have been neglecting an obvious duty if they had not made provision for the training of engineers, doctors, astronomers, mathematicians, geographers. . . . Vast territories were now to be organized and scientists and technicians had to be secured in a more systematic way" (Farrington, 1980: 195). Michael Grant divined that the museum's royal patrons "often had military purposes in mind" (Grant, 1982: 151). However, there is no evidence that any such training programs were carried on at the museum or that the state intended it to function as a military research institution. If the Hellenistic monarchs had useful knowledge in mind when they allocated expenditures to support libraries and the museum, they learned what their twentieth-century counterparts are relearning—support of theoretical investigations produces few socially beneficial results that can be counted as products of pure research.[5] Indeed, in the face of a meager return of useful knowledge the generosity of the Macedonian kings may have been short-lived. After one or two generations the productivity of Hellenistic science declined from the level of its initial upsurge (see Figure 4).

A common consequence of the allocation of state revenues for scientific research that then produces few practical benefits is resentment on the part of patrons and the public, and edgy defensiveness on the part of scientists. In the Hellenic period, even in the absence of public grants, that attitude had been reflected in the antiscientific plays of Aristophanes and in the ambiguous anecdotes surrounding the achievements of eminent savants. In the Hellenistic world similarly ambiguous traditions were preserved, intensified by resentment of the public support represented by the museum and the libraries. Euclid, when asked what good was the study of geometry, is said to have snapped: "Give him three obols, since he must needs make gain out of what he learns" (Bulmer-Thomas, 1971: 415).[6] Timon of Phlius (Timon the Skeptic; fl. 280 B.C.) described the museum as "a bird-cage, by way of ridiculing the philosophers who got their living there because they are fed like the choicest birds in a coop: 'Many there be that batten in populous Egypt, well-propped pedants who quarrel without end in the Muses' bird-cage'" (Athenaeus, 1927: 99 [ll. 22d–e]). But perhaps the most telling anecdotes surround the career of Archimedes (c. 287–212), who visited Alexandria, where he evidently studied with Euclid's disciples. Despite the fact that he was a mathematician and left behind only works on mathematics and theoretical mechanics, legends of technical wizardry have become attached to his name. A few have been confirmed, but some are clearly fabulous (Clagett, 1970: 213–14) and only reflect the hope that such impressive knowledge will somehow be turned to good use.

In any event, the research conducted in Alexandria became the jewel of Greek science. If Hellenistic science reflects the orientalization of the Greek tradition, the continued commitment of that science to abstract, discipline-oriented investigations reflects the Hellenization of an Oriental approach to higher learning. It was a symmetrical process, each tradition supplying its greatest strength and each overcoming its most disabling weakness—the Greek inability to provide institutional patronage of science and the Oriental bureaucratic compulsion to sponsor only directly useful knowledge. The vigorous hybridization of the Greek and Oriental traditions that occurred in Alexandria produced a golden age of science that has provided the model for all subsequent scientific cultures.

With cultures as with organisms, however, florescence must succumb to decline; and the decline of Hellenistic scientific learning is as relevant to the geography of science as are its growth and its maturity. The Alexandrian Museum lasted seven hundred years, almost as long as the European university. The Alexandrian Library was destroyed by Christian vigilantes in the fourth century; and in 415, with the murder by Christian fanatics of the pagan Hypatia (the first female mathematician), the museum came to an end (E. Kramer, 1972: 616; Sarton, 1959: 34, 156–57).

The causes of the decline of Hellenistic science remain uncertain. Too often these large-scale sociological movements are explained by mental or spiritual processes that are impossible to test. In one account, the question was posed: "What factors, social and intellectual, were responsible for the Greek scientific tradition losing its original character, and eventually being cut off entirely?" The answer given was: "Probably most important: during these centuries [i.e., after 100 B.C.] scientists gradually lost faith in themselves and in their methods of thought" (Toulmin & Goodfield, 1965: 129–30).

It is difficult to see how these spiritual afflictions can be regarded as fundamentally explanatory. Insofar as they enter the analysis at all they would appear to be symptoms, not causes. And why the malaise then, rather than sooner or later? Would it not be reasonable to look into factors external to science, the ecological and economic decay that blighted the social life of whole regions? Miletus, where it all began a thousand years before, declined in the Christian era—its harbor silted up. Syracuse, Archimedes' native city, failed to recover after it was sacked in 212 B.C. Rhodes declined in the next century. Pergamum, a center of learning that gave its name to "parchment," declined after 100 A.D. Alexandria survived, sustained by an indestructible agriculture; but Alexandrian science, which was no provincial affair and had been nourished by contacts with the central and eastern Mediterranean, was diminished when those contacts were broken. When cities of the Mediterranean came to ruin, Hellenistic science was afflicted.

Too little is known about the mechanics of the decline of ancient cities. But the spoliation of their agricultural bases was surely a factor. Deforestation of the Mediterranean region, which had already reduced the timber and energy reserves of many districts, was intensified during the Roman era, in order to build the navies of Rome and its enemies. Deforestation in turn intensified both soil erosion and silting of streams, as it always does. It is not the sacking of a city alone or the fracture of its political continuity that wounds it mortally. Sacked cities can be rebuilt, as they often have been, and their political life renewed, provided their agricultural hinterlands remain productive. Carthage was sacked by the Romans, but it recovered. When the city was finally abandoned it was the encroaching desert, not the raider's torch, that was the instrument of destruction. All over the ancient world, when the land itself was spoiled and its agricultural cadres disbanded, their homespun knowledge forgotten and their refined skills lost, dependent urban civilizations perished:

> Princes and lords may flourish, or may fade;
> A breath can make them, as a breath has made;
> But a bold peasantry, their country's pride,
> When once destroy'd, can never be supplied.
> (Goldsmith, *The Deserted Village*)

And with dying cities wither their fragile growths of science and learning.

DRANG NACH OSTEN

In response to the observation that he never ventured beyond the city limits of Athens, Socrates replied: "Very true, my good friend; and I hope that you will excuse me when you hear the reason, which is, that I am a lover of knowledge, and the men who dwell in the city are my teachers, and not the trees or the country" (Plato, 1892: 435 [*Phaedrus* 230–31]). Art flourished in late Paleolithic caves; music is a common feature of village life; but higher learning, as Socrates perceived, is intimately connected with cities. Being an ethical and political philosopher rather than a philosopher of nature, Socrates' interests were in "men" and the life of the polis, not in the world of nature. But his observation applies equally to scientific knowledge, for scientists, even naturalists and geologists whose research may be conducted in the countryside, interact with colleagues in urban centers and congregate in institutional settings that are usually located in cities. In classical Greece, science lacked supporting institutions; but the flow and exchange of ideas that cities naturally facilitate were especially favored in a society where, in contrast with the land-locked river valleys of the East, many cities were ports and centers of commerce. Conversely, science has failed to develop in social habitats where the city has been absent or only weakly developed—in the Paleolithic, in Neolithic farming societies, under pastoral nomadism, and, perhaps most revealingly, in European feudalism.

Lest it be thought that the location of scientific research in urban settings is an obvious and trivial association, it may be noted that technology, in contrast with science, was historically an affair of the country rather than the town. Mining and the construction of aqueducts and canals, which have been important engineering industries since antiquity, recruited engineers in the countryside; and the craft of millwright developed around rural water-power and wind-power sites in the European Middle Ages. The Roman bridges and aqueducts were not designed by urban intellectuals ensconced in institutions of higher learning; they were designed and built by village masons risen to the peak of their craft. In feudal Europe, when the country was ascendant over the town, technology flourished while science languished. Even during the Scientific and Industrial revolutions of the seventeenth and eighteenth centuries, science and technology progressed in alternating phases, and were separated not only by cognitive content and sociological status, but also along urban and rural lines. During those centuries, when some of the great figures of science, attached to the institutions of the capitals and university towns of Europe, were bending their thoughts to the possibility of extracting motive power from fire, the steam engine was in fact invented by Thomas Newcomen (1663–1729), an ironmonger innocent of scientific theory, working in the mining and industrial districts of Devonshire and the Midlands (Dorn, 1974). And, at about the same time, the economically significant process of smelting iron ore with coal instead of wood was hit upon by another ironmonger, Abraham Darby (1677–1717), at Coalbrookdale. The twentieth-century homogenization of science and technology has obscured the social, intellectual, and physical separateness of their historical developments.

The urbanocentric tendency of scientific development is related to the strong bonds between scientific research and institutional patronage. Except in its Hellenic expression, science has always maintained close ties with institutional sponsors, and cities in turn have always been the centers of institutional life. The pattern is clearly displayed in the period from the end of the Hellenic era (300 B.C.) to the rise of modern science in Northern Europe (1200 A.D.). During those fifteen hundred years "Western" science was preserved and nurtured under four systems of governance centered in the urbanized East—in the Hellenistic kingdoms, in the Byzantine (i.e., East Roman) Empire, in the Sassanid Empire of Persia, and in Islam—all of them in possession of major regions of intensified hydraulic agriculture and all of them displaying Oriental patterns of institutional patronage of higher learning. (See Figure 5.) In contrast, in the European West, where science was eventually to achieve its most luxuriant growth, agriculture, population increase, urban development, and scientific institutions and research were all retarded. Only with the development in the late Middle Ages of an effective technology of intensive

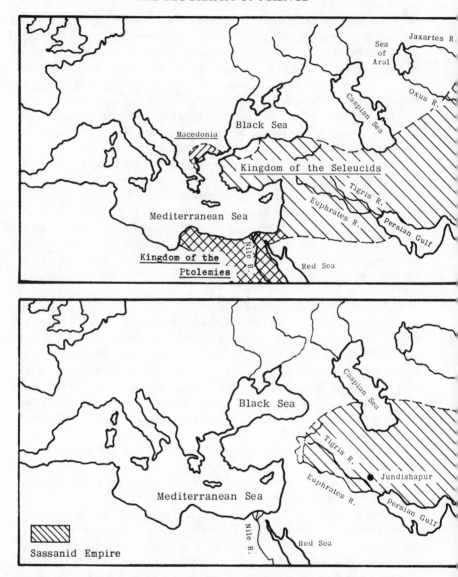

Fig. 5. Science was nurtured in four Eastern civilizations that intervened between classical Greece and medieval Europe, all of them in territories where hydraulic agriculture was well developed.
Top left, the Hellenistic kingdoms that formed after the death of Alexander the Great
Top right, the Byzantine Empire, which was founded in the fourth century A.D.
Bottom left, the Sassanid Empire of Persia (third to seventh centuries A.D.)
Bottom right, Islam, founded after the death of Muhammad (seventh century A.D.)

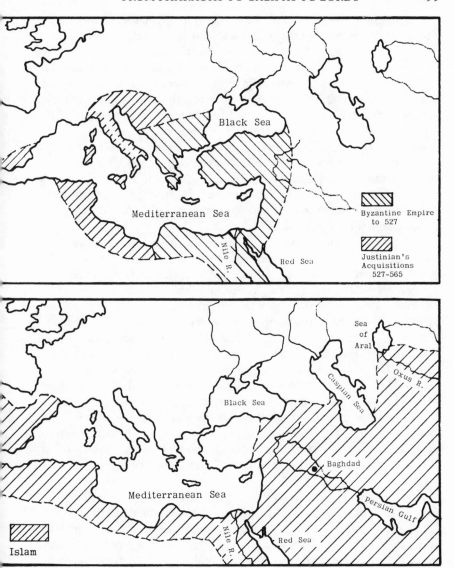

agriculture adapted to the unique conditions of transalpine Europe did cities, and with them scientific institutions, grow rapidly north of the fortieth parallel. It is yet another irony of the geography of culture that in the ancient world, and in the medieval world as well until the rise of the European universities in the thirteenth century, "Western" science was located mainly in the East.

During the ascendancy of the Roman Empire, which was a conglomerate of city-states, most of the cities under its control were east of Rome. In 100 A.D., of the twelve largest cities in the empire nine were in the East (Chandler & Fox, 1974: 303), and there were twice as many libraries in cities east of Italy as there were in Italy, Western Europe, and Carthage combined (Conzen, 1987: Map A-8). In the fourth century, the drive to the east of Roman civilization triumphed completely when Constantine the Great shifted the capital of the empire from Rome one thousand miles eastward to the entrance to the Black Sea and within easy reach of the vast agricultural resources of its western and northern coasts.[7] The new capital, Constantinople (on the site of ancient Byzantium), "stands at the crossing of two of the greatest trade routes of history" (Runciman, 1961: 12). Besides its access to southern Russia it commanded the richly productive agricultural regions of Macedonia, Thrace, and Egypt.

Constantine's simultaneous relocation of his capital to the East and his toleration of Christianity (and eventual conversion) make it especially difficult to unravel and assess the economic and political components of his motives. But there can be little doubt that the greater prosperity of the eastern empire over the western entered into his calculations. East and west displayed sharp contrasts in terms of demographic trends and economic prosperity. Between 200 and 600 A.D. the population of Italy declined by 50 percent from a peak of seven million (McEvedy & Jones, 1978: 107). Although in the east, too, some districts had been agriculturally and economically exhausted and were in decline, in many others population depletions were minimal and increases were common, indicating that overall the economic bases remained productive or had been reconstituted. Of the ten most populous cities in the Roman Empire at the time that Constantine moved the capital (330 A.D.), seven (excluding Constantinople) were east of Italy, six of them east of Greece (Chandler & Fox, 1974: 304).[8]

It was in these eastern territories, during this period, that the Byzantine Empire took hold and created an urban and institutional matrix in which science found a niche. With the surplus wealth of their region and with the highly centralized form of bureaucratic governance that Constantine established, Byzantine monarchs could build schools of higher learning, libraries "stocked with Greek manuscripts," hospitals, and, their most ostentatious creation, the splendid and monumental Hagia Sophia, built in the sixth century at an extravagant cost in gold (Runciman, 1961: 27, 257).

From the founding of Constantinople in the fourth century until the loss of Egypt to Islam three hundred years later, the Nile Valley was the granary of Byzantium and justifies characterizing it during that period as a hydraulic civilization. Wheat was the Byzantine Empire's primary agricultural product; and, according to A. H. M. Jones, the eminent historian of the early Byzantine Empire, "the paramount importance of agriculture in the economy of the empire can scarcely be exaggerated" (Jones, 1964, vol. 2: 769).[9]

With the loss of Egypt, and later of North Africa (eighth century) and Sicily (ninth century), the grain supply of the Byzantine Empire became precarious. Unfortunately for historical study, the authorities left unsatisfactory records of how they coped with the problem (Teall, 1959: 90). Two circumstances evidently enter into the account: the catastrophic plagues (bubonic and pneumonic) that struck Constantinople in the sixth century reduced its population and lessened the pressure on agricultural resources; and the empire leaned more heavily on the provinces it retained in Thrace, Macedonia, and Anatolia as well as on the possibilities of purchasing grain from southern Russia. In the absence of sufficient documentary evidence, the agricultural history of the Byzantine Empire (including the Anatolian plateau) after the seventh century will be fully clarified only through archaeological research.

The study of Byzantine science is in a scarcely more developed state than that of Byzantine agriculture. Many historical accounts of scientific learning between the fourth and seventh centuries A.D.—between the founding of Constantinople and the rise of Islam—fail to mention the fact that most of the scientists of the period worked under Byzantine governance. John Philoponus, perhaps the outstanding scientist of the sixth century, taught philosophy in Alexandria, which at the time was part of the Byzantine Empire. Yet he is sometimes discussed in the context of Roman science, presumably because he wrote in Latin (Stahl, 1962: 133), and sometimes in the context of Greek science, since the main thrust of his work was criticism of the physical and cosmological ideas of Aristotle (Clagett, 1963: 206–18), and only rarely as a Byzantine scientist (Taton, 1963: 443). Whether or not "Byzantine science" is a meaningful category remains to be settled by historical analysis. In terms of the internal, cognitive study of ancient and medieval science, it may be appropriate to erase the geographical boundaries of Byzantium and to emphasize the linguistic and philosophical traditions (Latin and Greek) to which Byzantine scientists belonged. However the historiography of scientific thought is compartmentalized, the physical realities of geography enter into the analysis of the social structure of science in the Byzantine world. To obscure completely the geographical provenance of science is to mask the economic, political, and institutional sources of its patronage—in this instance the schools, libraries, and hospitals in the cities of the eastern Mediterranean and of western Asia that were under Byzantine rule.

Although schools were established and supported in Constantinople and in several other cities under Byzantine governance (Taton [1963: 441] was probably mistaken in calling them universities) and a trickle of original research flowed from them, quantitatively and qualitatively Byzantine science was on the whole significantly inferior to Greek science. There had been nearly twice as many scientists between 600 and 200 B.C., the heyday of Greek science, as there were during the four centuries following the founding of Constantinople.[10] Most Byzantine scholarship consisted of compilations and translations of classical and Hellenistic texts, and it is generally agreed that few Byzantine scholars produced scientific works of outstanding originality.

While the "rise" of science is a common topic in historical studies, its enfeeblement is infrequently noted, much less studied. When it does come under review the physical environments in which scientific cultures falter tend to be overlooked in favor of explanations that stress spiritual exhaustion or spontaneous failure of intellect. Attempts to explain the low level of scientific productivity and originality in the Byzantine Empire have commonly attributed it to some form of ideological enervation in the face of an antagonistic ideology—as René Taton phrased it, "the total subordination of science to the Church" (Taton, 1963: 451). In the same spirit, Byzantine science and society have been compared to Soviet science, presumably in its least attractive, Lysenkoist manifestation:[11]

In the eyes of the Byzantine authorities, their state was the political embodiment of Christianity, much as, to the Russian authorities today [1965], the U.S.S.R. is the political embodiment of Marxism. This identification of the Christian Church with the Byzantine State tended to turn all intellectual debates into ideological ones: opinions were judged, not only on their merits, but also by their supposed political tendencies. Unorthodox intellectual views ran the risk of condemnation, not only as heretical but also as treasonable. The resulting intellectual atmosphere hardly encouraged the imaginative speculation needed if science was to make further progress. So natural philosophy was put in deep-freeze. (Toulmin & Goodfield, 1965: 154)

These conclusions lean too heavily on intellectual considerations alone and underestimate the influence of institutional development. Allowing for the fact that no Byzantine scientist was ever convicted of treason, the comparison is intended to show that whole scientific cultures are inhibited by the threat of political persecution whenever original ("unorthodox") ideas transgress the guidelines of a state religion. In that light, however, the role of scientific institutions tends to be obscured. It may, therefore, be instructive to scrutinize and criticize the argument. Was science in either society, Byzantine or Soviet, substantially suppressed by ideological persecution?

In the physical, natural, and exact sciences, where there is little occasion for ideological deviance, Soviet science, although heavily bent by institutional pressure towards applied research, is actively conducted, sometimes on a high level of originality. In some fields of mathematics Soviet research is renowned for its quality.[12] And, in general, wherever pure science is inexpensive and not explicitly doctrinal, the political authorities have remained ideologically indifferent to it.[13] Much of mathematics requires no costly facilities and ventures into no ideological disputes. And in the pure and esoteric field of bird-call research, a field that requires little more than tape recorders and that does not promote the (formerly) forbidden cults of dissidence, deviance, pornography, or violence, Soviet scientists are at the forefront (Boswall, 1987). Even in the field of physics, which, because of its economic and military potential, is inevitably the scene of more governmental intervention than bird-song studies, Soviet science has amassed a creditable record. Since 1958, when a Soviet scientist first received a Nobel Prize in physics, Soviet physicists have received as many Nobel awards as Britain and France combined. The relatively modest showing of Soviet science in the award of Nobel Prizes in general is at least partly the result of the prize founder's interests being inclined towards socially beneficial research (like his own invention of dynamite) and towards what he perceived to be the useful sciences. He endowed a prize for medical research, which, because it in large measure serves the individual, is not as strongly encouraged in the Soviet Union (where "public health" is favored) as in the Western democracies; but he created no prize for mathematics (which he evidently judged to be useless), in which Russians excel. On both scores the Soviet Nobel Prize record is diminished, but not because of any "risk of condemnation" or the suppression of allegedly dangerous scientific ideas.

Insofar as Soviet research is disabled by systemic conditions, they are likely to take the form of defects that generally have nothing to do with ideological repression—excessive teaching loads, endless meetings, limited access to copying machines, a shortage of computers (and even Petri dishes), and bureaucratic stultification—rather than any attempt to squelch ideas that undermine the "state religion." The notorious Lysenko affair was the expression of an aberrant excess of revolutionary zeal and an underdeveloped commitment to legal norms. It destroyed the science of genetics in the Soviet Union, but after twenty-five years it was repudiated and its most eminent victim is now commemorated by the Vavilov Institute in Moscow.

It seems similarly unlikely that in the Byzantine Empire repression was ever directed at science. It has often been suggested (e.g., Toulmin & Goodfield, 1965: 148) that the closing of Plato's old Academy in Athens in 529 by the emperor Justinian (r. 527–565) was an act that seriously damaged Byzantine science. In fact, however, when Justinian closed the Academy, which by then had become a center of Neoplatonic philosophy, his action was related to a program of dismissing pagan teachers by withholding the public funds out of which they were paid. Since the Academy in Athens was the only institution of higher learning in the Byzantine Empire that subsisted on its own endowment, a special law was required to bring it into line. But the repression was directed only against pagan intellectuals and only insofar as their teachings encroached upon the ruling theology. Neoplatonic ideas were favored by the pagan professors in Athens and had, moreover, a theological tinge which the authorities evidently held to represent a political threat. In Byzantine Alexandria pagan professors continued to teach, "partly because the Alexandrians concerned themselves largely (if not exclusively) with Aristotle, thus to some extent steering clear of the sinister religious and theosophical speculations of late neo-platonism" (Cameron, 1969: 8–9).

As a result of Justinian's action seven pagan teachers at the Platonic Academy lost their positions and emigrated to Sassanid Persia. Of the seven only one, Simplicius (b. ca. 500), may be regarded as a scientist. A few years later they migrated back to Byzantium, where they retained their convictions unmolested. Simplicius evidently returned to Athens, where he then composed the bulk of his works and may even have used the library at the "closed" Academy (Cameron, 1969: 22).[14] It will require more evidence than the closing of the Platonic Academy to prove that science faltered in Byzantium because of political repression. It was subversion, not science, that was the target of Justinian's hostility.

History records the martyrdom of many religious zealots, but few, if any, scientists. In the unlikely event that a scientist finds himself in a politically difficult position because of his research he can generally switch to a safer field of interest (as Galileo did after his persecution by the Inquisition); the religious zealot has only one agenda. It might even be expected that under conditions where political persecution is common science will flourish rather than falter, for intellectuals will then steer clear of risky political and philosophical studies and will be drawn to the sciences, many fields of which are politically and ideologically inert.[15] In Byzantium, as long as explicitly anti-Christian doctrines were avoided, study and research were tolerated, and in some fields encouraged. Aristotelian natural philosophy offered ample scope for Byzantine scholars in the fields of animal studies, theory of motion, mathematics, and astronomy. That they did not exercise themselves in these fields very energetically cannot, at least on present evidence, be attributed to repression. An isolated instance of governmental hostility directed at a single institution of higher learning cannot have sapped the vitality of a whole scientific culture over a period of a millennium. In seventy years of Soviet governance the Lysenko affair has not vitally damaged Soviet science; in the thousand-year history of the Byzantine Empire the closing of the Platonic Academy in Athens was proportionately less significant.[16]

A more likely explanation of the poor showing of Byzantine science is that it was institutionally undernourished, especially after the seventh-century loss of Egypt with its philosophical schools in Alexandria. Universities had not yet been created; observatories never found favor with Byzantine patrons; and the museum, in its Hellenistic manifestation as a research center, was of no interest to practical-minded monarchs who may well have known how little the Alexandrian Museum had produced in the way of practical results. Constantine and his successors were no novice pharaohs like Ptolemaios Soter. Insofar as Byzantine rulers favored scientific research they turned it in the direction of applications and technical improvements rather than towards imaginative speculation. They expected social utility from their civil service; and they had no interest in paying for research as the play of ideas or as the individual's craving for the ecstasy of discovery.

While Byzantine monarchs showed no interest in natural philosophy they retained veterinarians and they sponsored research in hippiatrics in order to maintain and improve their cavalry. A tell-tale difference between the civilization of Rome and that of Byzantium (probably related to ecological differences) is that the Roman soldier fought on foot, while the Byzantines commonly fought on horseback (Bivar, 1972: 273). Around the time of Constantine the army came to be based on cavalry, and the cavalry charge became the decisive action on the field of battle. In response to the new military technique Byzantine physicians, patronized by the monarchs, produced veterinary manuals, occasionally on a high level of originality:

The new demands of the military produced tracts on veterinary medicine . . . which formed major sources for the extant collection of veterinary materials known as the *Corpus hippiatricorum Graecorum.* . . . [Their] descriptions of equine disorders (glanders, for example) occasionally match and surpass anything before the nineteenth century. Recent studies . . . have detailed how innovative were these veterinary authors. (Scarborough, 1985: xi; see also Doyen-Higuet, 1985)

Institutional pressure cannot determine the theoretical results of research, but it can determine the field of study that will be favored.

A full social history of Byzantine science would display it in a more favorable light than does intellectual history alone. It would pay close attention to intellectually unambitious medical tracts, to treatises published by the veterinary surgeons retained by Byzantine monarchs, and to the many farmer's manuals and herbals that were produced under Byzantine governance. It is safe to conjecture that in a society where bureaucratic centralization was extreme the work of just such encyclopedists, translators, and writers of manuals on mundane subjects would be encouraged. And it is precisely the kind of work that historians intent on detecting theoretical novelty choose to ignore.

The enervation of theoretical science in Byzantium and the emphasis instead on the production of useful compendia could thus have been the result of the absence of congenial institutions that would encourage pure research, rather than of the repression of unorthodox ideas. In the same passage in which René Taton attributed the poverty of Byzantine science to the machinations of the Church, he added that, "paradoxically enough, they also paid attention to the practical problems arising in agricultural and veterinary matters, hospital services, the use of Greek fire, etc." And in the tenth century, "we know that the Byzantines busied themselves with the construction of hydraulic and other machines based on the ingenious ideas of Heron of Alexandria" (Taton, 1963: 451, 443). It was no paradox. The state favored technology and useful knowledge, not abstract theory. It was not that unorthodox scientific ideas were repressed by the threat of persecution; it was rather that avenues of research leading towards practical benefits were sponsored by an exceptionally pragmatic state while others were not. When the emphasis on applications becomes too insistent, patterns of patronage may deflect research from original speculative thought that promises practical benefits only sometime in the future (if at all).

An explanation in terms of the natural environment can admittedly do little more than theories of spiritual affliction or political repression in accounting for the history of Byzantine science. But, in accordance with the geographer's interest in space and place, including cultural as well as environmental considerations, a few differences between the eastern and western districts of the Roman Empire may be noted. However underdeveloped science was in general, it was much more robust in the eastern, Byzantine territories under conditions of thriving agriculture, population growth, and urbanization than it was in the western, Roman districts under conditions of a declining population and economy. There was much more activity and accomplishment, as measured by numbers of scientists, in the Greek East during the three hundred years after Constantine's conversion than there had been in the Latin West during the three hundred years prior to it. A count shows that there were nearly twice as many scientists in the East, even counting as western scientists those in the earlier, pre-Byzantine period who in fact worked in eastern cities.[17] Byzantine agriculture was sufficiently productive to sustain urbanization on a large scale along with the institutional developments that urban centers characteristically sponsor. And those institutions, some of which provided niches for scientific activity, inevitably fell under the domination of a state highly centralized by its military operations in a strategically precarious location.

There was in fact one characteristically Byzantine institution that in some measure sponsored science, albeit with a strong slant away from abstract thought and towards social benefits. The hospital, even today, is primarily a center of medical technology, not science. It dispenses therapies, remedies, and sometimes cures. But it may also be a scientific institution, insofar as it promotes theoretical research and sponsors the training and education of physicians along the lines of theoretical knowledge. In the urbanized centers of Byzantium hospitals provided science with its strongest institutional base during the period of late antiquity.

In a recent study of the founding of the hospital as an institution, Timothy Miller insists that the "true" hospital first came into being under Byzantine governance. Specifically, he argues that the older Roman *valetudinaria* failed to qualify as hospitals in that they "were designed to serve only restricted groups—the slaves of a particular estate or the troops of a given unit"; and the Greek *asklepieia* and *iatreia* fell short because they "did not normally provide their suppliants with a place to sleep, with food, or with nursing care" (Miller, 1985: 38–39). Moreover, he claims that these distinctions, which might appear to be excessively fine, are now commonly shared by institutional historians: "The great majority of researchers do now agree with the conclusion . . . that Christian churches of the Eastern Mediterranean opened the first centers of medical care sometime after the middle of the fourth century" (Miller, 1985: 30). Leaving aside the question of when the "true" hospital came into being, there is considerable evidence to support the conclusion that beginning in the fourth century A.D. in the East Roman (Byzantine) Empire, the hospital flourished both as an institution for the provision of medical services and also, to a lesser degree, as a center of medical science.

Constantine not only moved the capital of the empire to the more populous, more urbanized eastern territories, but he also instituted government support of charitable activities. Within a century of the founding of Constantinople, hospitals were opened in urban centers under Byzantine control; and by the seventh century well-developed institutions providing the full range of medical care and basing their procedures on the traditions of Galen and Hippocrates were established both in the capital and in other cosmopolitan centers of the eastern empire, including Jerusalem, Alexandria, Antinoupolis, Caesarea, and Antioch (Miller, 1985: 93–94). Although little is known of the organization of most of these institutions, some of them at least were comparable to the well-documented Pantokrator Xenon, which was founded in Constantinople in the twelfth century. That hospital, richly endowed by the emperor, provided fifty beds, nutritionally adequate meals for the patients, a staff of physicians and assistants, pharmacists, equipment, and service personnel. It was indeed owned by the emperor, who "had established it as a private, imperial foundation," and it was in many ways a typical institution of "the bureaucratic hospital medicine of the East Roman system" (Miller, 1985: 212, 223).

Christian mercy may have motivated the founding of these institutions, as Miller believes, but wealthy patrons and bureaucratic support made them possible: "Under the first Christian emperor, Constantine, the churches began to receive tax immunities and government grants to help carry out their charitable activities" (Miller, 1985: 21). Moreover, the Eastern habitat of the hospital suggests that Christian mercy was reinforced by the wealth of cities sustained by flourishing commerce and productive agriculture. It was in fact the combined effect of several factors that promoted the development of the Byzantine hospital—the social principles of Christianity, the largesse of government, church, and aristocratic patrons, the surpluses produced by trade and agriculture, and the concentration of population in the urbanized eastern Mediterranean and Asia Minor.

Although the urban focus of the medical profession was an ancient pattern, it was reinforced in the Byzantine world. It supports the conclusion that institutions, including those that sponsor science, flourish under conditions of urban development and demographic growth. Hospitals "came to occupy a significant place in the Byzantine conception of the polis," and they became "one of the essential amenities of city life" (Miller, 1985: 89, 10).

The utilitarian objectives of their patrons, combined with bureaucratic organization, were evidently enough to guarantee that the hospitals' resources were directed mainly at medical practice rather than theoretical research. Nonetheless, research was a component of this Pantokrator tradition. Some of the hospitals maintained libraries and teaching programs, and Galen and Hippocrates "remained the pillars of Greek medicine throughout the long history of the East Roman Empire." Byzantine hospitals preserved Greek texts in their libraries, and, in some measure, they also fostered new treatises and techniques (Miller, 1985: 35, 172–75).

Another institution, rivaling the hospital in the teaching of medicine, was church schools comparable to the monastic and cathedral schools that later served as sanctuaries for medical science in the Latin West. According to the historian of medicine Owsei Temkin, "in the realm of Constantinople, too, monasteries and ecclesiastical schools seem to have been the main teaching centers of medicine, but on a scale and with a spirit of learning that surpassed the West before the thirteenth century" (Temkin, 1962: 111). Byzantine science would surely make a better showing in institutional history than it has through the lens of intellectual history. As it has until now been developed, the history of Byzantine science leaves much to be explored in realms of research that historians of scientific ideas have studiously neglected.

The East Roman Empire is not the only civilization in Asia Minor during late antiquity that offers unexplored territory for the history of science. Byzantium was paralleled by another Eastern civilization, Sassanid Persia, in the heartland of ancient Mesopotamia. Sassanid Persia created a typical Oriental system of scientific institutions along with a typical Oriental economy based on hydraulic agriculture. The Sassanid dynasty was founded in 224 A.D.; and by the sixth century its royal residence, Jundishapur (see Fig. 5), northeast of present-day Basra, had become, according to George Sarton, "one of the greatest clearing-houses of philosophic and scientific ideas—Greek, Jewish, Christian, Syrian, Hindu, and Persian" and "the greatest intellectual center of the time" (Sarton, 1927: 417, 435). Since the Sassanid monarch, no less than his Byzantine peers, directed the science he patronized towards practical ends and social services, Jundishapur was not only a "great and flourishing city" (Chandler & Fox, 1974: 226) and a cosmopolitan center of intellectual life but also became the site of a famous hospital and medical school (Hau, 1979; E. Browne, 1983: 19–23): "Jundīshāpūr was especially important as a medical center; the medical teaching was essentially Greek, but with Hindu, Syrian, and Persian accretions. This medical school flourished until at least the end of the tenth century" (Sarton, 1927: 435).[18]

The line of scientific development and transmission from ancient Greece to modern Europe was drawn through a series of Middle Eastern cities—Alexandria, Pergamum, Constantinople, Jundishapur, and Baghdad—each of them supported on a productive agricultural base. The Sassanid monarch Chosroes (r. 531–579), who sheltered the Neoplatonist scholars expelled from the Platonic Academy by Justinian and in whose reign Jundishapur flourished as a scientific center, was also a water king. Sassanid Persia, situated on the Mesopotamian flood plain, achieved an unprecedented and unsurpassed development of irrigation agriculture in a region renowned as a homeland of hydraulic civilizations: "Increased population, the growth of urban centers, and expansion in the area of cultivation to its natural limits were linked in turn to an enlargement of the irrigation system on an unprecedented scale" (Jacobsen & Adams, 1958: 1256). The extent of cultivation and the number of inhabitants has never (to this day) been greater in the region than it was during the height of the Sassanid Empire in the sixth century. And Chosroes himself, whom George Sarton described as "one of the greatest among kings," was personally instrumental in the improvement of irrigation (Sarton, 1927: 436). A complex network of branch canals was constructed, the Tigris River was tapped by a large feeder canal nearly 200 miles long, and thousands of sluice gates were built "at a cost which could be met only with the full resources of a powerful and highly centralized state" (Jacobsen & Adams, 1958: 1256–57). The Sassanids built well, for even after the conquest of Persia by Islam, "the Sasanian investment in waterways and water control remained the fundamental basis of the economy" (Lapidus, 1981: 202).

Knowledge of the agricultural base of Sassanid Persia has been derived from an elaborate archaeological survey of the Diyala River basin east of the Tigris (Jacobsen & Adams, 1958). Comparable investigations might clarify the economic basis of Byzantine society and reveal the extent to which it, too, rested on the intensification of agriculture after the loss of Egypt. But, whatever the exact characteristics of their agricultural systems, the territories of the eastern Mediterranean and western Asia produced a sequence of centralized empires under Hellenistic, Byzantine, and Persian governance. In each of them centralization of authority produced scientific cultures that hybridized in some measure the institutional and intellectual traditions of the ancient Oriental kingdoms and classical Greece—state-dominated urban institutions in some of which scientific culture found a niche.

That geographical zone in the Middle East was to produce still another such civilization, this time under the aegis of Islam.

AGRICULTURE AND THE CULTURE OF SCIENCE

Contrary to the common impression that Islam rode onto the stage of history on horse- or camelback, in fact it strode onto the stage behind an ox-drawn plow. Like other peoples who have emerged from steppe or desert lands to establish conquest empires in habitats of settled agriculture, Arabs in history are not commonly looked upon as farmers. Yet, between the seventh and eleventh centuries, when Islam stretched from its western flank on the Atlantic coast of Portugal nearly to Calcutta in the East, and from Madagascar on its southern edge to the Pyrenees on its northwestern frontier and to Samarkand on its northeastern frontier, it initiated agricultural reforms so sweeping that, in Andrew Watson's detailed study of early Islamic agriculture, they have been described as an agricultural revolution (Watson, 1983: 123).[19]

Many new food and fiber crops of eastern or tropical provenance were adapted to the Mediterranean ecosystem and established in "nearly the whole of the early-Islamic world." Economically, the most important of these were rice, sorghum, sugar cane, cotton, watermelons, eggplants, spinach, colocasia, sour oranges, lemons, limes, bananas, and plantains. Hard wheat and artichokes, while indigenous to the Mediterranean, became widely diffused; and mangos and coconut palms were exotic plants that were adapted to the tropical regions of the Islamic world (Watson, 1983: 5; 1974: 9). This diversification of crops increased the intensification of agriculture along several lines: the growing season, which had traditionally been limited to the winter in the Mediterranean region, was extended to include the summer as well, since several of the new varieties could be adapted only to summer conditions; the new diversity of crops led to many additional combinations and systems of rotation, and thereby to an increased number of annual croppings; extension of the growing season and more numerous croppings employed idle land and labor; and increasing production inevitably required more water, "which in the lands of early Islam could be provided only by artificial irrigation" (Watson, 1974: 11).

The need for irrigation varied sharply from district to district, ranging from total dependence to the need only for occasional inundations to carry a crop through a dry spell or to take advantage of a flash flood. But all over the lands of Islam irrigation was practiced, and it always produced its characteristic centralizing effect. In many districts of the ancient granaries of Egypt and Mesopotamia, the hydraulic infrastructures had fallen into disrepair, and their reconstruction required the deployment of large labor forces (Watson, 1983: 104–7). In other districts new systems were built. And everywhere, canals and ditches had to be dredged and repaired constantly. Although the Islamic agricultural revolution was not confined to heavily irrigated areas, it was such regions that "may perhaps be regarded as the spearheads of agricultural advance" and the foci of centralized state authority.[20]

As everywhere in hydraulic societies, Islamic rulers often personally supervised the construction of irrigation systems (Rabie, 1981: 61); and where the monarch did not take a direct hand in construction, hydraulic projects were usually carried out under royal patronage and supervision: "Most new irrigation and reclamation projects were the work of the caliphs and of the caliphs' governors. . . . Other projects were sponsored by members of the royal family" (Lapidus, 1981: 189).

At the source of Islam, in the interior of Arabia, population had already been steadily increasing even before the time of Muhammad (570–632), and the pressure of population against desert and oasis resources was a major factor in the outpouring of the Bedouin armies that spread the teachings of the Prophet across the vast territories that became the Islamic world. In Arabia itself no further population increase was possible until the twentieth century, when the constraints of geography were overcome by the resources of geology; then population quadrupled. It was, instead, in the conquered territories, through the intensification of agriculture, that sharply increasing populations could be sustained:

The agricultural revolution was bound up with an ill-documented but none the less real demographic revolution which seems to have touched most parts of the Islamic world from roughly the beginning of the eighth till the end of the tenth century. Rising population levels and increasing levels of output of foodstuffs must continuously have interacted. (Watson, 1983: 129)

In the territories of ancient Mesopotamia, which became the heartland of Islam, population more than doubled in the eighth century; in the Maghreb (the "west," comprising Libya, Tunis, Algeria, and Morocco) it doubled from the seventh to the eleventh centuries; and in Spain a sharp decline during the pre-Islamic period was transformed into a steady increase during the Umayyad Amirate and the Caliphate of Córdoba (eighth through tenth centuries) (McEvedy & Jones, 1978: 151, 221, 101). Throughout Islam, agricultural hydraulicization resulted in demographic increase on an unprecedented scale: "The development of heavily irrigated agriculture in the early centuries of Islam allowed many regions to support a density of settlement which has not been equalled since— and sometimes had not been previously reached" (Watson, 1983: 111).

Increases in agricultural productivity and population in stratified societies generally produce increased urbanization, as surplus wealth is concentrated by an urbanized elite. In the Islamic conquest, old cities grew and new cities were built. Baghdad, on the site of ancient Ctesiphon, is said to have become ten times as populous as its predecessor (Lapidus, 1981: 181). Figures between 300,000 and 1,000,000 are generally given as estimates of its population, and they range as high as 1,100,000 (Chandler & Fox, 1974: 221; Russell [1958: 88–89] allows 300,000). Córdoba, on the Guadalquivir River in southwestern Spain, reached a population close to 1,000,000 under Islamic rule (Russell, 1958: 92); and several other Islamic cities had populations between 100,000 and 500,000 during a period when the largest European cities had populations numbered in the tens of thousands.[21]

In this urbanized habitat, where thriving cities generated and fostered high culture and a rich institutional life, Islamic science flourished. By the twelfth century Islamic Damascus had twelve *madaris*—institutions of higher learning (Chandler & Fox, 1974: 222); and similar institutions, along with hospitals, observatories, and libraries were founded all over the conquest.

Islamic science is renowned as a conduit through which Greek learning was "transmitted" to the West. That process was no doubt the effect, but surely not the intent. Indeed, the argument has been advanced that when the Abbasid caliphs first promoted the secular (i.e., *awâil*, pre-

Islamic, mainly Greek) sciences in the eighth century they were acting in their own interests no less than were the monarchs of the ancient Orient and late antiquity. Although Islam, like Byzantium, embraced a state religion, that condition did not foreclose the possibility of the exchange and comparison of ideas with the non-Islamic world. On the contrary, in the cities of Islam, Christians, Jews, and Muslims debated the tenets of religion, and since, prior to their acquisition of Greek learning, Muslims were often bested in these controversies the early caliphs encouraged secular learning in order to provide Muslim scholars with the logical and analytical means to defend the faith:

The interest of the caliphate in making Greek sciences available in Arabic most likely stems from this challenge . . . upon which the authority of the caliphate itself was based. It was therefore for the purposes of safeguarding the interests of the Muslim community that the early Abbasid caliphs turned the attention of scholars to the study of Greek philosophy and science. (Nasr, 1968: 70)

An alternate explanation of the initial impulse to appropriate Greek learning also attributes it to affairs of state and stresses its utilitarian objectives—the desire to master those fields, particularly medicine and astrology, that were deemed to be directly useful to the monarch. In the words of a medieval Islamic writer, "Now, of the servants essential to kings are the secretary, the poet, the astrologer, and the physician, with whom he can in no wise dispense. For the maintenance of the administration is by the secretary, the perpetuation of immortal renown by the poet, the ordering of affairs by the astrologer, and the health of the body by the physician" (Quoted in Sayili, 1981: 48). On either account Islamic science and its institutions would have had the support of the government, and the results were impressive (see Figure 6). The number of Islamic scientists during the four centuries after the time of the Prophet matched the number of Greek scientists during the four centuries following Thales.[22]

Reflecting its government support and institutional base, secular learning was especially well developed in the applied sciences—applied mathematics, astronomy insofar as it was applicable to astrology and to navigation, and those fields of study that were believed to be useful in agriculture and medicine. Original work was done in agronomy, botany, and pharmacology; and in some aspects of these sciences Islamic writers surpassed their Greek masters (Watson, 1974: 14; 1983: 145). Medieval Arabic mathematics is justly renowned, and while in part it earned its reputation in theoretical research, it consistently displayed a practical trend in its emphasis on arithmetic and algebra rather than on the formal theoretical geometry of the Greeks. Observational astronomy, which was directed at the compilation of "increasingly accurate tables for both as-

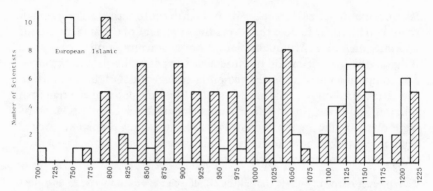

Fig. 6. Islamic science flourished with the support of royal patronage. In the large urban centers of Islam, libraries, schools, hopsitals, and especially observatories sustained the work of many scientists. During the period of Islamic ascendancy, Europe, with no large cities and low institutional density, had few scientists. (Ordinates based on a count of entries in the *DSB*.)

trological and nautical purposes" was equally practical (Crombie, 1959, vol. 1: 48–51); and Islamic medicine was derived directly from the medical institutions of Jundishapur.[23] Just as Ptolemaios Soter had recruited Athenian scholars to staff his newly founded museum in Alexandria, the early caliphs gradually transferred the leading physicians of Jundishapur to their new institutions in Baghdad.

As in the ancient Oriental kingdoms, the utilitarian bias of Islamic science obliterated the distinction between rational and occult studies, for both were regarded as potentially useful in the service of society and the state. Astrology, which was valued for "the ordering of affairs [of state]," was systematically studied and, more than pure astronomy, was often substantially patronized by the caliphs (Sayili, 1981: 47–49). It was the pervasive and venerable confidence in the utility and benefactions of higher learning, and not any interest in transcendental mysteries, that enabled astrology to enjoy the caliphs' favor.

The full extent of the state's sponsorship of research can only be seen in the social development of Islamic science—in its "universities," its hospitals, its libraries, and especially, its observatories. University-like institutions (*madaris;* sing. *madrasa*) were founded in many cities from Granada to Samarkand. Leaving aside claims that the *madaris* were fully comparable to the universities that later arose in Western Europe, they were undeniably schools of higher learning that were supported by the state. They were, moreover, institutions that, albeit to a limited extent, fostered the study of science.[24]

As a center of scientific work, perhaps the most important Islamic institution of higher learning was the celebrated House of Wisdom (*Bayt al Hikma*), founded in Baghdad in the ninth century. Greek scientific manuscripts were collected from Byzantine sources and brought to the House of Wisdom where scholar-translators undertook the Herculean task of translating into Arabic the philosophical and scientific tradition of the ancient world. A measure of the effort expended on translating Greek texts is that, even now, more Aristotelian writings—the works of Aristotle and his Greek commentators—are available in Arabic than in any European language (Nasr, 1968: 70). Another measure of scholarly activity in general is that by the tenth century Islamic Córdoba had 70 libraries, one containing between 400,000 and 500,000 manuscripts.

The fabled libraries of Islam, established by the caliphs, reached from Spain to Central Asia. In tenth-century Shiraz (in southwestern Iran) a library was founded by the monarch 'Adud al-Dawla; and the famous scholar Ibn Sina (980–1037), known in the West as Avicenna, worked in the royal library in Bukhara and left an account of its impressive quality:

I found there many rooms filled with books which were arranged in cases, row upon row. One room was allotted to works on Arabic philology and poetry, another to jurisprudence and so forth, the books on each particular science having a room to themselves. I inspected the catalogue of ancient Greek authors and looked for the books which I required; I saw in this collection books of which few people have heard even the names, and which I myself have never seen either before or since. (Quoted in Halsell, 1990: 12)

Royal libraries were also established in Basra, Mosul, Rayy, Merv, and Kharizm (Hitti, 1961: 270–71; Hoberman, 1987: 5). An anecdote testifies to the high value that Islamic scholars placed on books. In a tenth-century Islamic study of the etiology of melancholy, the author cited deep tragedy as one of its possible causes, as from "the loss of a beloved child, or of an irreplaceable library" (Quoted in Critchley, 1987: 5).

But, the love of learning alone could not have accounted for those libraries. The formation of collections with holdings of tens of thousands and hundreds of thousands of manuscripts was clearly dependent on the willingness of a caliph or, perhaps, a wealthy aristocrat to underwrite the costs. However, even that would not have been enough if the costs had remained beyond reach. As a result of an eighth-century Muslim victory over Chinese forces east of Samarkand (near present-day Dzambul in the Kazakh S.S.R.), the Chinese technique of papermaking was acquired by Islam, and paper soon replaced papyrus from Egypt as well as parchment, both costly items: "The advent of paper lowered the price of books so drastically that public and private libraries soon became common throughout the Islamic world. Schools—and bookshops—began to pro-liferate" (Hoberman, 1987: 4).

In addition to libraries, Baghdad became the site of the first Islamic hospital; and as the Islamic agricultural revolution took hold and "large cities grew by leaps and bounds" hospitals and medical centers were established throughout the conquest—in Damascus, Cairo, Tunisia, Granada—supported mainly by the government. Some of those hospitals contained libraries and lecture halls and became centers of medical teaching and research (Hamarneh, 1962: 381, 370–74). Unlike the Christian schools in Syria, where medical education was apparently held in low esteem and medical students were segregated in order to prevent profane learning from corrupting sacred studies (Vööbis, 1965: 285–86), Islamic medicine and hospitals were under secular control and were held in high regard (Hamarneh, 1962: 379).

Schools, libraries, and hospitals all promoted scientific research directly or indirectly. The most distinctive Islamic scientific institution, however, was the astronomical observatory. Despite questionable references to "observatories" at the Alexandrian Museum and in Jundishapur, it seems likely that the astronomical observatory was as much an Islamic innovation as the hospital was Byzantine (Sayili, 1981: 346–56). Observatories were constructed in several cities of the eastern caliphate, generally under the patronage of the state. The Abbasid caliph al-Mamun (786–833), who had founded the House of Wisdom and patronized scientists who were familiar with Greek learning, also built observatories in Baghdad and in Damascus. During the thirteenth century a large observatory was constructed in Marâgha, in a fertile region of Azerbaijan (Saliba, 1987). It is said to have contained a library of 400,000 volumes and "it apparently incorporated a veritable school of astronomy and the awâil [secular] sciences in general" (Sayili, 1981: 194, 219). In fifteenth-century Samarkand, sustained by irrigated orchards, gardens, and cropland, the celebrated scholar-prince Ulugh Bey (1393–1449), grandson of Tamerlane, founded a madrasa and an observatory, which were to serve as important links to European science. There were still other Islamic observatories, and all of them depended on the patronage of rulers, many of whom were personally interested in astronomical or, rather in most cases, astrological research. The importance that Islamic astronomers attached to the precision of their observations necessitated the use of exceptionally large instruments; and these in turn, along with the observatory structures, the staffs of astronomers and support personnel, and their affiliated libraries, entailed costs so high that they could only be met through government support.[25] Through its observatories medieval Islam established a tradition of observational science that was unequalled until the achievements of European science in the fifteenth and sixteenth centuries.

The objectives of the caliphs in patronizing astronomical research were manifold, but they were primarily focused on the interests of state-craft, including the requirements of religion.[26] Accurately determining the times of prayer and especially of community prayers required the services of astronomers who, accordingly, were often attached to the staffs of mosques (Sayili, 1981: 24). It was also necessary for religious reasons to ascertain the direction of Mecca from every place of worship; and the same astronomical geography that enabled worshipers to face Mecca during prayers produced navigational techniques that were of use both to seamen and to desert travelers. But the strongest single motive behind royal patronage of astronomy was the putative divinatory power of astrology. Despite its occasional condemnation by religious authorities on the grounds that it misdirected piety towards the stars rather than God, it remained the "most popular" of the secular sciences and "thrived well in the royal courts." Horoscopic astrology for the benefit of individuals contributed little, if anything, to the development of science. It was rather "the royal patronage of astrology [that] resulted in the erection of important observatories and furthered the study of pure astronomy and the mathematical sciences" (Sayili, 1981: 47, 419). Even the fatal contradiction, inherent in all divinitory sciences, between the predictability of a strictly determined future and the consequent impossibility of altering that future, failed to quench the interest of caliphs. For, however unpleasant and unalterable the future may be, it was argued, it is nonetheless more reassuring to know it than to have it come as a surprise.[27]

The institutional density of Islamic science accounts for some of its achievements and characteristics. Scholars and scientists were on the staffs of schools, libraries, mosques, hospitals, and especially observatories, where teams of astronomers and mathematicians were employed. One result of the opportunities and support that these institutions offered scientists was the remarkable upsurge in scientific activity, as measured by the number of Islamic scientists. Another was a characteristic research profile, identifiable since its first display in the ancient Oriental kingdoms, which exaggerated utility, public service, and the interests of statecraft.

The geography of Islamic civilization defined the habitat in which Islamic science flourished and from which it passed the torch of learning to Europe. Islamic science was a component of a sociological configuration that included a set of interlocked developments: an agricultural revolution based on a diversification of crops, an intensification of agricultural production through the reconstruction and extension of irrigation systems, rapid population growth, urban concentration, and opulently wealthy monarchs who lavished encouragement, goodwill, and money on a variety of cultural institutions.

How is this Islamic social formation to be classified? Although temporally it overlapped European feudalism, Islam is unique among ancient and medieval societies in almost never being characterized by historians as feudal. In 1950 at Princeton University a "Conference on Feudalism" was held, the aim of which was "to test the extent of repetition in history." Feudalism was discussed in the contexts of Western Europe, Japan, China, Mesopotamia and Iran, ancient Egypt, India, Byzantium, and Russia—but not Islam.[28]

Spatially, Islam and European feudalism were sharply disjoint (albeit contiguous)—Islam essentially confined to the semi-arid, hydraulic regions of the Middle East, central Asia, Egypt, the Maghreb, and Spain; and feudalism, for the most part, to the moist, potentially fertile plains of Northern Europe. Moreover, if "setting the stage for capitalism" is taken to be a diagnostic of feudalism (as it often is; e.g., Levenson, 1956: 570, and Keddie, 1981: 765), Islam is again excluded from the category of feudal society. For, although capitalism has made inroads into Islamic regions it has done so only belatedly, in the nineteenth and twentieth centuries under the economically motivated tutelage of the advanced capitalist societies of Europe and North America. Conversely, the heartland of European feudalism, which was to be transformed into the heartland of capitalism, was never penetrated by Islamic civilization. On both counts, as well as on grounds of their ecological differences, Islam and feudalism appear to have been mutually exclusive. The *Cambridge History of Islam* has only one reference to feudalism (Holt, Lambton, & Lewis, 1970: 536), indirectly confirming the conclusion that "the [medieval] Middle East as a whole cannot be helpfully classified as 'feudal.'" (Keddie, 1981: 766).

Marxist analysis is curiously taciturn about an Islamic mode of production of any description. Indeed, Marxist historical principles suggest a connection, not between Islam and feudalism, but rather between Islam and the Asiatic Mode of Production, both of which seem to have *inhibited* the development of capitalism. Each fostered highly centralized societies and was heavily dependent on hydraulic agriculture; and both have proven to be exceptionally insular and durable, long failing to nurture either the feudal or the capitalist social formations that were adopted in Europe.

In science, too, Islam and the Asiatic Mode of Production displayed marked similarities, with research situated in courts, palaces, and cultural institutions sponsored by caliphs, kings, emperors, and pharaohs, and inclined towards useful applications including the prognosticatory fields of knowledge. The Islamic observatories, in particular, concentrated in the eastern caliphate and almost nonexistent in Spain and the Maghreb, suggest an easy acceptance by Islam of the traditions of the scientific cultures of the ancient East.[29]

In one respect, however, the scientific cultures of Islam and of the ancient civilizations of Egypt and Mesopotamia were inevitably different. Islam, like all societies that have received Greek science and philosophy, reflected, along with a dominant emphasis on useful knowledge cultivated in institutional settings, an interest in pure research, undirected towards any governmental or public benefits. Every civilization that has inherited Greek intellectual traditions, however heavily centralized and institutionalized that society may have been, has hybridized in some significant measure the unfettered inquiry of individual scholars with the constraints and agendas, but also the support, of scientific institutions. While astrology was promoted for reasons of state, the same patronage that nurtured it also enabled astronomers, following the lead of Greek natural philosophers, to study "the science of motion of the heavenly bodies, and to overcome its inconsistencies, so as to attain perfection of knowledge" (Nasr, 1968: 147). In Islam the inheritance of a firm Hellenic tradition sustained a fruitful activity of abstract study alongside the pursuit of directly useful knowledge. This combination of the intellectual spontaneity of the scientist and the institutional program of his patron has shaped the development of science ever since the Hellenic Greeks added the philosophy of nature to the lists and recipes of the ancient Oriental scientist-bureaucrats.

In those agricultural and industrial societies that have embodied the traditions of both the ancient East and ancient Greece, the personal scientific interests of individuals have competed with the interests of the public communicated through institutions. In ancient Egypt and Mesopotamia, individual interests received no social reinforcement, and hence are only slightly in evidence. In the Hellenic world, in contrast, it was the institutional support of science that was lacking; individual savants, if they were to study at all, were compelled to do so on their own in pursuit of their own intellectual interests. The pattern of combining the two missions first became apparent in Alexandria, where Hellenic science was orientalized by being settled in an institutional environment while retaining its heavy emphasis on abstract research. In the scientific institutions of Islam the dual pattern continued but the balance tilted back, as institutional objectives imposed by monarchs outweighed the quest "to attain perfection of knowledge." In the next stage, in European civilization, which was in environmental terms the negation of Islamic society, the balance was to tilt once again towards the Hellenic tradition of natural philosophy.

THE EMPTY QUARTER

Agriculture in the lands of the Mediterranean and the Middle East is an exceptionally fragile enterprise. Throughout most of its range, unless the fields are constantly tended, the land reverts to desert. In contrast, Northern Europe offers vast tracts of level and fertile fields that, through the concurrence of heat and light provided by moist summers, have the potential to produce large surpluses and retain their fertility even when neglected and abused. Fernand Braudel summarized the divergent ecologies of the Mediterranean and the North European agricultural zones:

[In the Mediterranean zone] the thin layers of topsoil, which only the modest wooden swing-plough can scratch, are at the mercy of the wind or the flood waters. They are enabled to survive only by man's constant effort. Given these conditions, if the peasants' vigilance should be distracted during long periods of unrest, not only the peasantry but also the productive soil will be destroyed. During the disturbances of the Thirty Years' War, the German peasantry was decimated, but the land remained and with it the possibility of renewal. Here lay the superiority of the North. In the Mediterranean the soil dies if it is not protected by crops; the desert lies in wait for arable land and never lets go. (Braudel, 1975, vol. 1: 243)[30]

However, the "superiority of the North" did not assert itself until the late Middle Ages, and then only because it was facilitated by an agricultural revolution as sweeping as that which had enabled Islam to stretch from the Atlantic coast to Samarkand. Prior to that revolution Northern Europe was a veritable "empty quarter," a *Rub' al-Khali*. High civilization had been confined to those regions of the world where agricultural intensification and massed peasantries, or the easy opportunity to trade with or dominate such regions, as in the case of Greece, made food surpluses available for sustaining urban civilization. During the millennia when urbanized societies flourished all around the world in more southerly latitudes, Northern Europe lay sequestered in village life, indifferent to the allures of high civilization. Then, while Islam conquered, Northern Europe began to stir; but the physical and cultural differences between the two civilizations were only slowly reduced. In the first half of the tenth century, when Islamic Córdoba had nearly a million inhabitants and shared with Constantinople the distinction of being one of the two most populous cities in the world, Paris had a population of 38,000 (Nawwab, Speers, & Hoye, 1980: 66). Even four hundred years later London may have reached only 100,000.[31] In the ninth and tenth centuries, the total population of Islam (including the Maghreb) was twice as large as that of northwest Europe. (Today the populations of the same territories are in the ratio of 3 to 4.) While Córdoba's main library had over 400,000 volumes, the monastic libraries of Christian Europe had holdings counted in the hundreds. And the disparity of the scientific enterprises of the two civilizations was almost as large, perhaps as much as ten times more scientific activity occurring in Islam than in the Latin West.[32] The caliph, in command of an extended territory of intensified agriculture, had the resources and the authority to create notable cultural institutions; the manorial lord, superintendent of a local farming system that was only beginning to realize its productive potential, had neither.

Over half a century ago, the Belgian historian Henri Pirenne argued that the Moslem occupation of the Mediterranean coasts (essentially the Mediterranean agricultural zone) drove Western civilization both inland and inward upon itself (Pirenne, 1937: 2–5). At various times during its conquest, Islam was in possession not only of Syria, Lebanon, Egypt, the Maghreb, and Spain—that is, the eastern, southern, and western shores of the Mediterranean—but also the southwestern coast of France and the islands of Sicily, Malta, Sardinia, Crete, and Rhodes.[33] (Greece may not have been worth conquering.) What had been "our sea" (*mare nostrum*) for the Roman world was now "their sea" for Western civilization. North of its lost world Europe reposed in a somnolent state—territories marked by low population density, little urban concentration, the absence of intensive agriculture, the dominance of country over town, low levels of political centralization, negligible institutional density, and essentially no science. Until it produced its own agricultural revolution, Northern Europe waited in abeyance.

CHAPTER FIVE

=

ABOVE THE FORTIETH PARALLEL

NEW TRENDS IN CULTIVATION

Except for Europe, none of the world's civilizations—in the Near East, Egypt, Greece and Rome, India, Ceylon, Southeast Asia, the Far East, or in the Americas—originated north of the fortieth parallel of latitude. Only European civilization has developed so far from the equator—from Rome, on the forty-first parallel, to the sixtieth parallel, which runs north of Scotland and passes through Oslo, Stockholm, Helsinki, and Leningrad. Indeed, in that northerly belt of the Old World, only the European region could have supported an agricultural economy and an urbanized way of life. The remainder of those northern territories are for the most part tundra, waterless desert beyond the reach of even the most ingenious irrigation technology, or steppe suitable for nomadic herding. It was only exceptional environmental conditions—the moderating effects of the Gulf Stream, the concurrence of moisture and warmth during the European summer (in the Mediterranean, the rain falls in the winter), and immense stretches of level land—that made Europe a candidate for agricultural civilization. In its unique geographical habitat, Europe, although it arrived last among the world's civilizations, eventually created an extraordinarily productive agricultural system; and, unlike all other high civilizations (except Greece), it did so largely on the basis of rainfall farming, without the benefits (or costs) of hydraulic agriculture.

The history of European science and material life is bound together with a sequence of developments that ultimately induced the nation-states of Europe to become a major source of scientific patronage. These developments included an agricultural revolution that resulted in the production of large surpluses, the consequent ascendancy of the city over the country and the concomitant founding of the European university as a unique institution, and a military revolution that served, in lieu of irrigation engineering, as a politically centralizing force.

Prior to the twelfth century, science was a negligible activity in Northern Europe. With little surplus wealth to concentrate in urban centers, without centralized governments that might have patronized higher learning, and lacking bustling ports and centers of commerce comparable to those of ancient Greece, science failed to develop, either under institutional auspices or as the activity of leisured individuals. The European agricultural revolution transformed the situation. Wealth accumulated rapidly, cities flourished, and the university came into being. By the middle of the fourteenth century European science was thriving, so much so that a sharp debate has divided historians over the question of whether "modern" science originated with the Catholic intellectuals of the fourteenth century or only in the sixteenth century with Copernicus and Vesalius.

Under European ecological conditions, with agriculture dependent mainly on rainfall, the centralization of authority that eventually led to lavish state support of scientific institutions and research could not have sprung from the hydraulicization of agriculture as it had in the East. On the contrary, rainfall agriculture neither required nor permitted the intervention of the state. The rain fell on all, whether they paid their taxes or not; water for crops could not be allocated by any earthly authority; and no agricultural public works were required that would call the state into action. Instead, the intensification of European agriculture was abetted by quite another set of physical conditions.

Nearly thirty years ago Lynn White, Jr., formulated the thesis that a European agricultural revolution occurred between the sixth and ninth centuries, overlooked by historians but historically consequential:

Nowhere are the urban roots of the word "civilization" more evident than in the neglect which historians have lavished upon the rustic and his works and days. While the peasant has normally been a lively and enterprising fellow, quite unlike the tragic caricature of combined brutishness and abused virtue presented in Millet's and Markham's "Man with the Hoe," he has seldom been literate. Not only histories but documents in general were produced by social groups which took the peasant and his labours largely for granted. Therefore while our libraries groan with data on the ownership of land, there is an astonishing dearth of information about the various, and often changing, methods of cultivation which made the land worth owning. (White, 1962a: 39)

The main physical problem that had retarded the development of European agriculture was the moist, heavy, clayey soil, which the Mediterranean scratch plow drawn by a pair of oxen was too feeble to work. White described the introduction of the heavy, wheeled plow with coulter, plowshare, and moldboard, which tore up the earth at the root line and turned it over, forming a ridge and a furrow. This behemoth of wood and iron assaulted the earth with unprecedented violence and was resisted by immense friction, which in turn could be overcome only by even greater traction. The traditional ox, or even the yoked pair, was no longer adequate. Teams of four, six, and eight oxen were now required to provide the necessary pulling power. The productivity of plowing was further increased by improved harnessing and especially by the inventions of the horse collar and the horseshoe. Harnessed with a collar (instead of a yoke) and shod with iron, the horse, with its greater speed and endurance, replaced the ox as the puller of plows and wagons. (By transferring the pressure point from an animal's windpipe to its shoulders, the horse collar allowed a four- to fivefold increase in traction.)

Two more interlocked innovations completed the agricultural revolution of the Middle Ages: the practice of three-field rotation, which increased productivity by fifty percent, and the concomitant introduction of a spring planting (made possible in Northern Europe by spring and summer rain), which resulted in a large increase in the growing and consumption of vegetables, thereby improving the health of both the farmer and his soil.[1]

The type of social system that developed around this constellation of innovations suggests that European feudalism can be understood as an ecological and technological system—agricultural intensification without either hydraulicization or public works in a region of low population density and no large cities. It was in these ways the opposite of the Asiatic Mode of Production. Lynn White seems to have been thinking along this line when he stated that "the manor as a co-operative agricultural community was, in fact, typical not of the Mediterranean lands but only of areas which employed the heavy plough, and there appeared to be a causal connection between plough and manor" (White, 1962a: 44). It may indeed have been the newly introduced heavy plow, rather than the ancient hand mill, that in Marx's memorable aphorism "gives you the feudal lord."[2]

These, in any event, were the innovative features of the medieval revolution in European agriculture: an increased use of iron in the form of the horseshoe and agricultural implements, the heavy plow in place of the traditional Mediterranean scratch plow, horse-power in place of ox-power, a three-field triennial rotation of crops, and a large increase in the planting and eating of vegetables. The social consequences of these innovations were comparable to the predictable consequences of the intensification of agriculture anywhere—rapid population growth, urbanization as measured by large increases in the number and size of cities, monumental architecture in the form of the soaring Gothic cathedrals, and an increase in institutional density. A component of that last development was the emergence of a novel institution, the university.

TOWN AND GOWN

Since the origin of civilization in the ancient East, higher learning had been sponsored by a variety of institutions—by the Egyptian House of Life, by scribal schools in Egypt and Mesopotamia, by the Astronomical Bureau in China, by the Alexandrian Museum and Library, by schools under Byzantine governance in Constantinople, Antioch, Alexandria, Berytus, Athens, and Gaza, and by the Islamic libraries, House of Wisdom, and university-like *madaris*—all of them founded and patronized by the state. But the European university was unique:

This formation of an institutional shell around the business of learning was unique to Europe. This pattern did not exist in the Asiatic world, nor had it existed in Byzantium or in the Arabic world, where institutions of higher education depended on a prince or an emperor. The least successful of the medieval Italian universities were those closely dependent on a sovereign. The University of Naples was never very distinguished; the University of the Roman Couria was worst of all, because it was too close to the pope. (Hyde, 1988: 19)

As wealth accumulated and European cities grew, educational institutions and the higher learning they sponsored developed in step with them and produced what has been called the "renaissance of the twelfth century." The medieval university generally came into being in burgeoning centers of commerce and urban life, just as the pace of economic life was quickening.[3] Over the next three hundred years more than eighty additional universities were founded across Europe in a generally northward drift (Figure 7).

Authorities on the history of the university have affirmed the connection between the awakening of learning and the urbanization and economic invigoration of Northern Europe: "Late in the eleventh century, and closely connected with the revival of trade and town life, came a revival of law, foreshadowing the renaissance of the century which followed" (Haskins, 1957: 7). Hastings Rashdall, the historian of the medieval university, referring to the Parisian career of Peter Abelard (1079–1142), who "inaugurated the intellectual movement out of which [universities] eventually sprang," observed that it "coincided with the first steps in the rapid rise to commercial and political importance of the ancient stronghold of the Counts of Paris" (Rashdall, 1936, vol. 1: 43, 62). More recent studies have supported these conclusions:

The city of Paris and its university grew up together. It was exactly in the same era—from the early years of the twelfth century through the first decades of the thirteenth—that Paris became the greatest city in northern Europe, the capital of an increasingly powerful French kingdom, a major commercial center, and the site

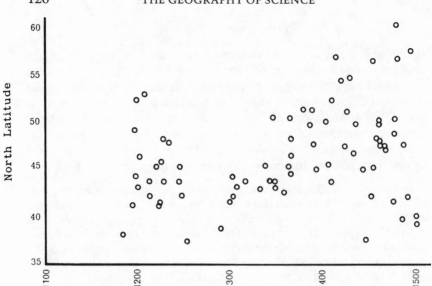

Fig. 7. From the twelfth through the fifteenth centuries the European universities were founded in a generally northward drift. Note that the vertical axis is calibrated as degrees north latitude. (See Shepherd, 1929: 100.)

of what was arguably the first and—certainly for a long time remained—the preeminent university. (Ferruolo, 1988: 22)[4]

It was not some ill-defined economic growth that accompanied the emergence of the medieval university, but rather the specific development of intensive agriculture, producing the food surpluses that could sustain growing urban populations, which included corporations of students and masters. Writing of the founding of the first European universities at a precise moment in history—"ten years on either side of the year 1200"— J. K. Hyde observed:

This would have been inconceivable had there not been the great growth of towns and cities in Europe, beginning at the end of the tenth century, gathering pace gradually through the eleventh, and becoming a breakaway rise in the twelfth and thirteenth. . . . The spontaneous universities of Italy were formed in large, growing cities, distinguished, I think, by fat agricultural regions, with a food surplus which meant that relatively cheap living was possible for an idle student population. (Hyde, 1988: 14)

That "idle" population was no exclusive cohort of monastic initiates or palace retainers. The European university was a commercial marketplace of ideas open to anyone who paid the fees. And most of the early universities, both ecclesiastical and civic, were independent of any central political authority. In that unique institutional environment, science, in the form of mathematics, astronomy, medicine, and natural philosophy, found a niche and eventually evolved into a scientific culture of unmatched vigor and originality.

Despite its freedom from the authority of a state, the medieval university was, like all institutions, under constant pressure to serve society. In feudal Europe, with its decentralized agricultural economy, there was no caliphal or pharaonic authority to organize higher education. Instead, it fell largely to church institutions—monasteries and cathedrals—to provide the setting in which learning was fostered.[5] Under these conditions education was inevitably embedded in an ecclesiastical curriculum. But even prior to the twelfth century, secular studies were allowed a place in the curricula of the monastic schools insofar as those studies were judged to serve sacred or practical interests. The seven liberal arts, which formed the basis of education in the early universities, consisted of grammar, rhetoric, logic, arithmetic, geometry, astronomy, and music, all of them secular fields of study and in some measure calculated either directly or indirectly to serve utilitarian purposes. Arithmetic and astronomy were introduced "chiefly because they taught the means of finding Easter"; and, in a further nod to secular knowledge with a utilitarian bent, geometry was taught in a markedly practical manner, presenting Euclid's propositions but omitting the demonstrations (Rashdall, 1936, vol. 1: 34–36). The more abstract studies of theology and philosophy were taught on a higher level only after the *trivium* and *quadrivium* had been mastered.

In discussions of medieval learning too much is sometimes made of its scholastic intellectualism and its philosophical disputatiousness. The reality, however, was and is that in an institutional setting there is inevitably an expectation, even a requirement (albeit sometimes disregarded by both staff and higher authorities), that practical objectives be kept in mind and properly addressed. The result, not infrequently, is substantial tension between the lofty principles that are enunciated and the earthy tasks that are pursued. This ambiguity makes it particularly difficult to demystify institutions and reach a clear understanding of their missions and of the means by which their patrons attempt to ensure that those missions are accepted and accomplished.

Nonetheless, in the case of the medieval university some of the evidence is suggestive of a socially utilitarian agenda. A century after their first founding, the universities were growing, not primarily as centers of intellectual achievement or ecclesiastical interest, but rather in response to a need for lawyers and administrators: "From a more practical point of view their greatest service to mankind was simply this, that they placed the administration of human affairs—in short, the government of the world—in the hands of educated men" (Rashdall, 1936, vol. 3: 456–57).

At the University of Paris the trend was evident even earlier:

It has often been noted that in the decades when the university was being formed, the scholarship of the Paris masters, especially the theologians, became less speculative and creative, more practical and routine, than before. This intellectual change has usually . . . been attributed to the needs of the growing numbers of students who came to the schools seeking training for careers that would be spent in the administrative ranks of the church or secular rulers. (Ferruolo, 1988: 37)

Although many students directed their studies at the canon law and in their careers served the church, many others followed the paths of commerce and government service. And even when graduates were employed by the church, the university was serving secular and practical interests insofar as the church performed social services.

The European university and the Islamic *madrasa* shared several traits, and the similarities have raised questions over whether Europe is indebted to Islam for its institutional organization of science to the same extent that European learning is indebted to Islam for transmitting Greek philosophical and scientific texts.[6] In both institutions, theology was the supreme subject of study, at least in the eyes of the authorities who sponsored them. But in both, law and teaching were in fact the professions for which students were most commonly prepared. And, although some of the teachers at both institutions distinguished themselves for their learning and originality, neither institution was designed as a center of research. The most significant difference between them was sociological: while the *madrasa* was established and patronized by the state, in the form of the caliphate, medieval Europe lacked a state authority sufficiently centralized to perform such a function. In Europe universities were commonly organized by guilds of students and masters, often under the jurisdiction or auspices of the church. But even on that score the university and the *madrasa* cannot be categorically separated, for as Europe flourished economically, the state became increasingly centralized and absolute and contested the church for control of the universities. Although it was only in the nineteenth century that the states of Europe began to allocate substantial funds for the support of universities, they showed an early interest in the public services that the universities might perform. Indeed, at Oxford, from early in its history, the state competed with the church for control, one of its interests being the supply of bureaucrats for the civil services (Rothblatt, 1988: 12).

The medieval church and state played their roles with sufficient ambiguity to provide an opportunity for abstract learning, including the sciences, to consolidate a place in the curricula of the universities. Logic was a field of Aristotelian philosophy that appeared to be useful in theological and legal studies as well as ideologically neutral (as it had likewise been perceived to be in Islam). In the monastic schools, prior to the founding of the universities, logic had already become a sanctuary for abstract thought and the secular intellect:

Here the teacher was untrammelled by the lurking uneasiness of conscience which haunted the medieval monk who loved his Virgil: there was nothing pagan about syllogisms: the rules of right reasoning were the same for Christian and for pagan alike, and were (as was thought) essential for the right comprehension and inculcation of Christian truth. Under cover of this idea teacher and pupil alike were enabled in the study of dialectic, and perhaps in dialectic only, to enjoy something of the pleasure of knowledge for its own sake. (Rashdall, 1936, vol. 1: 37)

Relatively free of institutional constraints, logic became the pure science of the early medieval curriculum. It gave the university an Aristotelian accent that it has retained ever since, and it thereby facilitated the securing of a place in the university for Aristotelian science in the form of natural philosophy.[7]

At the Universities of Oxford and Paris in the thirteenth century, logic not only flourished but even predominated among the seven liberal arts (Kibre & Siraisi, 1978: 127–28); and the following century saw the florescence of natural philosophy in the same institutional centers. The degree of anarchy and ambiguity that was often tolerated by institutional administrations, sometimes knowingly, and occasionally reinforced by the benign or impotent indifference of the patronizing authorities, allowed pure science to gain a foothold in the university, and at times to climb to great heights. From the renaissance of the twelfth century to a peak in the fourteenth century the scientific enterprise in Europe displayed an unprecedented level of vigor, surpassing even the Islamic achievement. Then, with the massive depopulation brought on by the bubonic plague, and with the disruptions in France of the Hundred Years' War, scientific activity in Europe faltered for more than a century. It revived at the end of the fifteenth century, and since then its growth (interrupted for only a brief period in the seventeenth century) has far outpaced that of any other scientific culture. (See Figure 8.)

When the plague struck Europe in the mid-fourteenth century an immediate consequence for science and learning was the loss of productive scholars. Although the lives of scientists were not recorded attentively in fourteenth-century Europe, many scientists were drawn from the higher clergy, a cohort about which much more is known and which suffered a mortality of over 35 percent (Campbell, 1931: 136). Secondary effects of the pestilence, including subsequent outbreaks, were as consequential as the direct loss of scholars following the initial epidemic, and stretched the havoc into succeeding generations. The number of university stu-

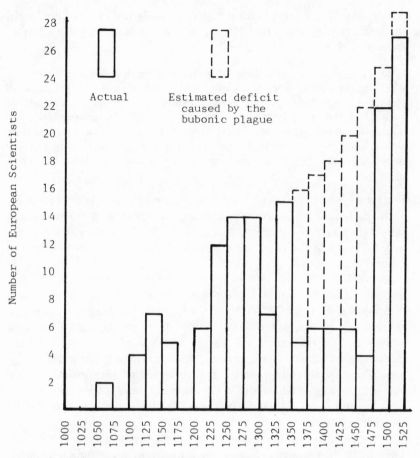

Fig. 8. The Black Death, which struck Europe in the middle of the fourteenth century, is estimated to have resulted in the death of one-third of the population. It also depressed scientific activity for more than one hundred years. Data represented in the graph suggest that, if not for the plague, European science would have developed at a steady pace, and there would be less reason to postulate a break between "medieval" and "modern" science. (Ordinates based on a count of entries in the *DSB*.)

dents was sharply depressed, and the doleful effects on elementary and secondary education, although poorly chronicled, were evidently severe (Campbell, 1931: 145, 174ff.). The damage to preuniversity education depressed scientific activity a generation later. In combination, these consequences of the pestilence produced cultural devastation that reverberated through European civilization until the eighteenth century (Thorndike, 1926–1927).

All over Europe the universities were stunned, and they recovered only slowly. Fifty years after the first outbreak the University of Paris "was suspended on account of the 'mortality'" (Campbell, 1931: 97). At Oxford the flourishing research of the first half of the fourteenth century came to an abrupt and melancholy halt. In concluding a recent history of the early period of the University of Oxford, J. A. Weisheipl stated:

This chapter closes with the devastating plague of 1349, which carried off most of the scholars we have been concerned with. It is most likely that William of Ockham died in Munich that year. Thomas Bradwardine ["perhaps the greatest English mathematician of the fourteenth century" (Boyer, 1959: 66)], consecrated archbishop of Canterbury at Avignon on 10 July 1349, died of the plague at Lambeth on 26 August. That year saw the deaths also of Thomas Buckingham, John Dumbleton, William Sutton and countless other Oxford scholars. On the most conservative estimate, one out of every three persons in England died of the plague that year. With it came the end of a unique period in the history of the University of Oxford (Weisheipl, 1984: 658).

The scientific depression caused by the plague intervened between what have been classified as "medieval" and "modern" science. Whether such a disjunction is enlightening remains an unresolved controversy. (For a review of the issues, along with the opinion that there was no intellectual discontinuity between "medieval" and "modern" science, see Crombie, 1969.) Viewed in geographical perspective, the history of European science would show few distinctions between a "medieval" and a "modern" tradition. Instead, it would be seen as a single developing culture in a well-defined physical habitat, beginning with an agricultural revolution, urban concentration, and the rise of the university in the twelfth century. Although "modern" science, which literally means nothing more than "recent" or "not ancient" science, may be a convenient improvisation, it is an artifact of intellectual history, its plausibility heightened by the effects of a demographic discontinuity. Geographically, European science is a coherent entity on the same analytical level as the scientific cultures that preceded it—in Islam, Persia, Byzantium, the Hellenistic kingdoms, classical Greece, and the Asiatic societies of the East. The division of European science into intellectual movements and the designation of a "modern" achievement may stroke the European ego, but its historiography loses sight of the essential unity of the European scientific enterprise.

GUNPOWDER AND GOVERNMENT

Eventually, regions of Europe coalesced into nation-states under increasingly centralized governments, and science was shaped into an instrument of statecraft and public service no less firmly than it had been in the ancient monarchies. However, before that recapitulation of the institutional patterns of the East could be completed, centralizing forces would have to concentrate political authority. In the semi-arid East hydraulic agriculture had encouraged political centralization; in early modern Europe comparable political and institutional patterns were promoted by a new technological development—the management of an innovative system of warfare that evolved between the fifteenth and eighteenth centuries and that a growing number of historians, following the lead of Michael Roberts, have termed a "military revolution."

In the fifteenth century, gunpowder and firearms began to play a decisive role on the battlefields of Europe, and by the end of the century they had transformed the politics and sociology of war. The "gunpowder revolution" undermined the military roles of both the feudal knight and the feudal lord and replaced them with enormously expensive gunpowder armies and navies financed by central governments. It was not simply that the armored knight was defenseless against the new ordnance. In fact, knights continued to proffer their services and to maintain retinues of archers and pikemen. It was rather the economic dimension of the revolution in warfare that neither knights nor noblemen could master, for the new artillery lay beyond the means of any individual captain or lord and could be financed only by royal treasuries. At the beginning of the Hundred Years' War (1337–1453), the primary means of combat were still the longbow, the crossbow, the pike, and the armored knight mounted on an armored charger. At the end, gunpowder artillery carried the day.[8] Cannon rapidly developed from wrought iron barrel-tubes to immense castings weighing thousands of pounds, and their costs inflated proportionately.[9]

The new ordnance required large increases in the budgets of European governments. During the second half of the fifteenth century, as the military revolution took hold, "central government tax revenues in western Europe doubled in real, per capita terms" (McNeill, 1982: 105). Infantry, now bristling with handguns, became once again a dominant arm on the field of battle. Over the next two centuries the armies of several European states increased tenfold in numbers, and during the last seventy years of the seventeenth century alone the French army grew from 150,000 to 400,000 soldiers.[10]

As ascendancy in warfare passed from the defense, in the form of the medieval castle, to the offense, in the form of gunfire that could shatter castle walls in a few hours, European governments poured money into the development of new systems of fortification. As costly as were guns and powder, the new fortresses—the *trace italienne*—were even more costly. Building earthen ramparts studded with angled masonry bastions, from which to confront the attackers with numerous guns of its own, strained the resources of even the richest European states (McNeill, 1982: 90).

Sometimes the costs were ruinously high and produced consequences beyond the field of combat into the world of politics:

The scheme to surround Rome with a belt of eighteen powerful bastions was abandoned in 1542 when the construction of one bastion alone was found to have cost 44,000 ducats (about £10,000). Palmanova, built on the Venetian frontier in Friuli in the 1590s, was originally intended to have twelve bastions; but this was soon reduced to nine on the grounds of cost. It was a wise decision: the Republic of Sienna had lost its independence forty years earlier because its leaders embarked upon a programme of fortification that they could not afford. (G. Parker, 1988: 12)

Offense and defense alternated in a pattern of challenge and response. Relentlessly, costs mounted, and warfare became the province of centralized monarchies. After the Hundred Years' War the "kingdom of France . . . emerged on the map of Europe . . . centralized as never before and capable of maintaining a standing professional army of about 25,000 men" (McNeill, 1982: 83). Moreover, the military revolution was not confined to land warfare. At the same time that firearms were changing the conduct of war and the structure of society, the transformation of the warship from the galley to the galleon was further intensifying political centralization by placing additional financial burdens on the states of Europe. The galley went the way of the armored knight and was replaced by heavy, ocean-going sailing ships armed with guns firing through gunports close to the water line (to keep the center of gravity low for the sake of seaworthiness). The new warship had to be strong enough to withstand both the recoil of its own guns and the incoming fire of enemy guns. The privately owned armed merchantman was first supplemented and then replaced by royal navies, which voraciously demanded an unending flow of tax revenues. And these revenue requirements, already inflated by the costs of land armies and fortresses, frequently resulted in royal bankruptcies and even contributed to revolution.[11]

Although the military revolution was centered in Europe, its consequences were worldwide. Despite their staggering costs, the new weaponry and the systems of defense they called into being were steadily developed, and they moved Europe from the wings to the center of the world-historical stage. The centroid of European civilization shifted decisively from the Mediterranean to the states of northwest Europe as the gunned ship made the Atlantic ports the springboards that launched Europe on its domination of the coasts of Asia and the Americas.[12]

From the fifteenth century onward, the creation of these national armies and royal navies resulted in political centralization as inevitably as had the hydraulic projects of the ancient and medieval East. Arsenals and shipyards were maintained as state-owned and -controlled public works comparable to the dams and canals of hydraulic societies. And military conscription, foreshadowed in the Swedish army of Gustavus Adolphus in the seventeenth century and established as a national principle by the French Revolution, became the corvée of modern Europe. If the central governments of the ancient Oriental kingdoms derived their authority from the vital services they performed in the organization of hydraulic agriculture, the increasingly centralized state in early modern Europe derived much of its authority from organizing and financing the new armies and navies that firearms necessitated. Subsequently, the orchestration of the three centuries of overseas imperialism that was sustained by the gunpowder revolution provided the European state with still another centralizing function to perform.

As the idea of a military revolution in early modern Europe has caught on among historians, they have directed attention to its unambiguous effects in concentrating political authority in strong central governments. Michael Roberts referred to "the new principle of concentrating military power under the absolute control of the sovereign," and stated that "the transformation in the scale of war led inevitably to an increase in the authority of the state. . . . Only the state, now, could supply the administrative, technical and financial resources required for large-scale hostilities. And the state was concerned to make its military monopoly absolute" (Roberts, 1967a: 210, 204–5). William McNeill observed that "new weaponry began to favor larger states and more powerful monarchs," and referred to "the centralizing effect of the new technology of war" (McNeill, 1982: 80, 95). Michael Duffy concluded that

to supply the expanded armed forces involved considerable state intervention in the economy. . . . The state became the biggest single purchaser of food, clothing and metalware, and the biggest shipbuilder in maritime areas; it became the biggest single source of employment for the construction industry in its demand for dockyards, for barracks and for fortifications—particularly this latter since the entire system of fortifications of all towns and strongpoints in Europe had to be modernized to the bastioned-trace system. . . . Armed forces and state power . . . were inseparably connected in early modern Europe. (Duffy, 1980: 6–7)

And Paul Kennedy observed that

the changes in military techniques which permitted the great rise in the size of armies and the almost simultaneous evolution of large-scale naval conflict placed enormous new pressures upon the organized societies of the West. Each belligerent had to learn how to create a satisfactory administrative structure to meet the "military revolution"; and, of equal importance, it also had to devise new means of paying for the spiraling costs of war. (Kennedy, 1987: 56)

The new historiography has shaken the old assurance, abetted by a dogmatic Marxist emphasis on the importance of property relations, that socioeconomic factors alone, without their technical components, determined the direction of social change in early modern Europe. William McNeill summarized this revision of historical thought:

Marxists unfortunately share the nineteenth-century Eurocentric blinkers that inevitably limited Karl Marx's vision of human history. Among Europeans of his age, the supremacy of the market and of the pecuniary nexus seemed assured for all time—past, present, and future. From the perspective of the late twentieth century this no longer seems a self-evident truth, and historians may therefore soon become sensitive to the military-technical and political aspects of the rise of European capitalism. (McNeill, 1982: 116)

It now appears that European societies became politically centralized in their own way without the structures and devices of hydraulic agriculture. A historically unique military technology deflected European societies towards centralized authority and introduced a dynamic social force that has relentlessly favored technical development. In the technological basis of its political structures, just as in its geographical environment, it was Europe that was deviant. The East was the norm—the torpid East, where public works, centered around the requirements of hydraulic agriculture, were technologically conservative in keeping with the traditions and risks of farming. In Europe, the new military technology was inherently innovative. In a complete reversal of the traditions that had characterized hydraulic civilizations, the risks among European states were highest for those who failed to change. These technological imperatives led to a reversed pattern of development—the East had been "stagnant"; Europe was dynamic. European monarchs constantly demanded more destructive and more mobile artillery, lighter handguns, swifter ships, and stronger fortresses.

Ultimately, in their need for innovation, they appealed for scientific services and the application of scientific enlightenment. However, applied science in Europe is largely a nineteenth-century development, and until the merger of theoretical science and industrial technology after 1850 the quest of European governments for the fruits of higher learning followed a circuitous route. In a curious historical development, the niche that applied science currently occupies in industrial societies was already filled in early modern Europe by the occult sciences—alchemy, astrology, and magic. Before any enlightenment could be applied to the solution of military and economic problems, occult learning had to be dislodged from its favored status.

FROM MAGIC TO APPLIED SCIENCE

European science perpetuated a tradition of occult learning that had originated in the ancient East. It eventually lost favor in Europe, not with the rise of modern science, but only in the eighteenth century, when the pace of industrialization quickened and it became increasingly clear to the governments of Europe that the kind of improvements and benefits they valued could be provided only by craftsmen and by the services of scientific institutions, not by diviners and soothsayers.

Beyond the fields of medicine, mathematics, and calendrical astronomy, science had traditionally been called upon to benefit the state and society mainly through occult studies—those that might command either physical processes, like alchemy, or cosmic forces, like magic, or might predict the future, like astrology. Confidence in the practical effectiveness of arcane knowledge and mysterious processes had been sustained by long records of proven success. Had not the alchemy of irrigation transformed desert into El Dorado? Had not inscrutable metallurgical procedures in fiery crucibles transformed useless ores into valuable metals? Was not the influence of heavenly events on earthly affairs an undeniable daily occurrence of vital importance to sailors and farmers? And did not the close study of the celestial realm enable savants to predict the seasons? Ever since the beginnings of civilization in the Oriental kingdoms, it was an element of conventional wisdom that knowledge in general, but often in the realms of what are now seen as occult learning, could be useful in securing goods, success, status, health, and power. Among the learned, occult wisdom was embodied in an intellectual tradition that survived intact and that intermittently flourished until the European Enlightenment.

At the time of the founding of the universities, magic was already a central feature of the scientific tradition in Europe. A. C. Crombie, in his survey of medieval science, stated that "the preoccupation with the magical and astrological properties of natural objects was, with the search for moral symbols, the chief characteristic of the scientific outlook of Western Christendom before the 13th century" (Crombie, 1959, vol. 1: 16–17). Throughout the Middle Ages and into the seventeenth century, astrology and medicine were closely related, and astrologers sometimes held professorships in the newly founded universities. In Italy, beginning in the thirteenth century, "the university centers acted as magnets to draw together all those concerned with Aristotelian natural philosophy, astronomy and astrology, and medicine." Logic, astrology, and natural philosophy "were regarded as particularly important preparatory studies for physicians," and, hence, students had "a very practical incentive to study astrology (including astronomy) and Aristotelian natural philosophy, since these subjects served as preparation for the prestigious and lucrative medical profession" (Kibre & Siraisi, 1978: 138–39, 135).[14] As late as the end of the seventeenth century, the "century of genius," chemistry and alchemy were scarcely distinguishable, and Isaac Newton's extensive interest in alchemy, which outweighed his interest in physics and mathematics in the time and effort he devoted to it, has led his most informed biographer to ask, "Have we perhaps mistaken the thrust of Newton's career?" (Westfall, 1975: 196. See also Westfall's definitive biography of Newton [1980: esp. ch. 9] and Dobbs, 1975).

Among government and church authorities, the same bias towards the occult sciences was evident, since occult studies promised to produce the useful knowledge that pure science, or, in its historical manifestation, natural philosophy and pure mathematics, had conspicuously neglected. One of the objectives of the Latin West in acquiring Muslim learning was to "share this rumoured magical power," and one of the consequences of the acquisition was that the occult sciences "fruited tropically" in Europe just as they had in Islam (Crombie, 1959, vol. 1: 52). In the thirteenth century, Roger Bacon (c. 1219–c. 1292) laid great stress on astrology as a component of mathematics in the service of the theological authorities who had become patrons of education and learning in Christian Europe. He maintained that astrology might strengthen faith, predict the destruction of Islam, and warn of the coming of the Antichrist (Crombie & North, 1970: 382). (He also valued geography, a less occult science, on the grounds that it might *locate* the Antichrist.)

Accounts of the occult tradition generally recognize its social role and its reputation as a source of practical benefits. The close connections among science, magic, government, and the church during the medieval and early modern eras have been noted by Keith Thomas in his authoritative history of the occult sciences. In England, "it was not uncommon for medieval kings to receive astrological advice. . . . Astrology was primarily the concern of the Court, nobility and Church." For the sixteenth and seventeenth centuries, "the famous Elizabethan [astrological] practitioner, John Dee, was no back-alley quack, but the confidant of the Queen and her ministers." And Thomas quoted a seventeenth-century "astrological writer" to the effect that "astrology had not 'much conversed at any time with the mean and vulgar sort, but . . . hath been ever most familiar with great personages, princes, kings, and emperors' " (Thomas, 1971: 301–2). While Newton was immersed in his studies, both rational and occult, astrology was avidly patronized by public figures from the king down: "Astrologers' advice was sought along the entire political spectrum, from Charles II to leading Levellers and Ranters" (Curry, 1987a: 245). Quantitative data for the early modern era are sparse, but it appears that between 1350 and 1400 nearly half of European astrologers were employees of secular or ecclesiastical dignitaries (Campbell, 1931: 127).

As an intellectual tradition, the occult sciences rested heavily on the "Hermetic Books" of the legendary ancient Egyptian soothsayer, Hermes

Trismegistus (Hermes, the Thrice Blessed), works that, until the seventeenth century, were commonly believed to have originated in Mosaic times (Dannenfeldt, 1972: 306).[15] The entire accretion of occult learning and practices—the magic, philosophy, witchcraft, astrology, and alchemy—and the Hermetic writings in particular, sustained an intellectual tradition, often influencing the thought of many scholars who contributed to the development of the rational sciences—a cohort that included, among many other scientists of the sixteenth and seventeenth centuries, Copernicus, Tycho Brahe, Kepler, Bacon, Gilbert, and Newton.

Those mystery studies and practices also embodied principles and procedures that specifically fostered an interest in applied science and technology. Magic, among its arcane and transcendental doctrines, postulated forces pulsating everywhere; forces in which all bodies are immersed and that animate the universe, which would otherwise be lifeless matter; forces that control the physical world and that adepts, steeped in the mysteries of the ancient theology, can learn to command. Thus, side by side with its philosophical system, magic offered a method whereby knowledge of the deep secrets of nature could be used to master physical processes, not unlike what has come to be expected of the applied sciences.

Indeed, among its patrons magic was valued more as a technology than as a science, as a system for commanding powers and predicting the future, rather than for understanding the world. Frances Yates, whose historical studies have legitimated magic as one of the sources of modern science, stated that "the Hermetic movement thus encouraged some of the genuine applied sciences, including mechanics" (Yates, 1967: 259).[16] And Paolo Rossi, in his study of Francis Bacon (1561–1626), characterized the attitude that merged science, occult learning, and the crafts as a "view . . . common to all Renaissance writers concerned with the significance of magic" (Rossi, 1968: 19). Magic thus displayed the same dual nature as science. Among its enthusiasts it may have been a system of abstract thought; to its patrons it was a source of powers.

For clarifying the relationship between science and society in early modern Europe the major significance of the occult tradition lies in its sociological implications and institutional affiliations. Insofar as sovereigns continued to believe in the operative powers of astrology, alchemy, magic, and hermeticism, and regarded them as instruments of statecraft, those fields of study provided a rationale for the institutional development not only of occult learning itself but also of the rational sciences to which they were akin. In the sixteenth century, Tycho Brahe (1546–1601) established astronomical observatories under royal patronage, and astrological knowledge was among the objectives of the investigations conducted in them. Although astrology was surely of greater

interest to his patrons than to Brahe himself, he was nonetheless atten-
tive to its requirements in conducting his research. King Frederick II, the
Danish monarch who subsidized Brahe's work, had a keen interest in
weather prediction and asked Brahe to compose a treatise on the subject,
an effort that astrology was expected to serve. On another occasion, in
deference to the king's wish to have him present a series of lectures at the
university in Copenhagen, Brahe delivered an "Oration on Astrology"
(Dreyer, 1963: ch. 4).[17] Like scientists today, sixteenth-century astronomers, including Brahe, sometimes won support for their work by appealing to their patrons' interests in what was regarded as useful knowledge.
Astrology often failed, they claimed, only because of the inadequacy of
existing star charts, thereby conveniently justifying additional support
for their own research (Sayili, 1981: 43–44).[18]

Brahe's disciple, Johannes Kepler (1571–1630), was employed as an
astrologer by European monarchs and by the imperial general, Albrecht
von Wallenstein. At the time of the supernova of 1604, the emperor,
Rudolph II, consulted Kepler on the astrological significance of the event;
and two years later Kepler published De stella nova, which he announced
as "a book full of astronomical, physical, metaphysical, meteorological
and astrological discussions, glorious and unusual" (quoted in Gingerich,
1973: 297). Kepler was unquestionably much more devoted to planetary
theory and his own metaphysical interpretations of astronomy than he
was to astrology, but matters were reversed with the patrons who endorsed and supported his research.

The prevalence of magic until the eighteenth century, both among
European scientists and in the reward systems of their patrons, is now
generally recognized by historians. The converse process, the decline of
magic, has proven to be a more difficult historical problem. Keith
Thomas described it as the "most difficult problem in the study of magical beliefs," and he confessed that "it is one which the historian of Tudor
and Stuart England must . . . find peculiarly intractable" (Thomas, 1971:
643). Intellectual historians may be inclined to attribute the decline of
magic to its alleged suppression by the Scientific Revolution or the
"Newtonian system" (Thomas, 1971:352), on the assumption that occult
learning and rational learning are incompatible. But Newton's own interest in alchemy immediately calls that contention into question. Moreover, his theory of comets has been shown to have been an attempt to
rescue astrology from quackery and recover it as a scientific discipline,
not to condemn it as a scientific heresy:

Newtonian cometography transformed astrology. . . . It changed the practice
which should deal with comets from popular divination to theologically oriented
natural philosophy, giving comets a profound but scarcely less dramatic function

and prophetic meaning. They would cause the Deluge, terminate and restore life on Earth and rejuvenate the Sun and the stars. (Schaffer, 1987: 243)

If there was thus no conflict between the occult and the rational in Newton's mind, historians should think twice before postulating such a conflict in general.

The tendency to see magic strictly in the context of the history of ideas has bedeviled attempts to explain both its durability and its decline. In searching for an explanation of decline, Keith Thomas first focused attention on magic as an intellectual system; and despite its close affiliation with science, he saw its demise as having resulted from "a series of intellectual combats" between science and magic that magic allegedly lost: "The triumph of the mechanical philosophy [i.e., the new science of the seventeenth century] meant the end of the animistic conception of the universe which had constituted the basic rationale for magical thinking" (Thomas, 1971: 644). Recognizing the inadequacy of his own suggestion that magic was defeated by science, Thomas conceded that an explanation confined to intellectual history was unsound: "One cannot simply attribute the [decline of magic] to the scientific revolution. There were too many 'rationalists' before, too many believers afterwards, for so simple an explanation to be plausible" (Thomas, 1971: 647).[19]

Instead of stressing the intellectual status of magic, Thomas then suggested that magical beliefs be considered "in their social context," specifically in connection with the development of technology. Through its promise to command physical processes, magic was competitive with the crafts and technique. It is thus more plausible to seek the decline of magic in the increasing success of technology rather than science. As Europeans mastered their environment through technical innovation, so the argument goes, they could dispense with the rigamaroles of the magi: "The decline of magic coincided with a marked improvement in the extent to which [the] environment became amenable to control. In several important respects the material conditions of life took a turn for the better during the later seventeenth century" (Thomas, 1971: 650).

Whatever its cause, the decline of magic as a practical art severed a link between learning and society that science, if only it could be made useful, might be expected to restore. But, while the governments of Europe may have known unambiguously what they had wanted from the occult sciences, they were less certain about what they might gain in the form of practical benefits from the rational sciences that they hesitantly began to patronize in the second half of the seventeenth century. However much their largesse is draped in altruism, patrons expect benefits, either personally or to "society" as they define it, in return for their philanthropy. The new science was accompanied by considerable propaganda about its potential for service to mankind. But the monarchs of Europe remained ambivalently both hopeful and skeptical, and before they would lend their full financial support they insisted upon more than a promissory note.

One of the most revealing indications of governmental ambivalence towards the new science occurred in the 1660s at the time of the founding of the Royal Society of London. In the seventeenth century, as the number of scientists rapidly outgrew the number of university professorships, scientists increasingly sought to indulge and share their common interests through memberships in scientific societies, academies, and clubs, in lieu of faculty affiliations. By the middle of the century groups of scientists in various parts of Europe had begun to assemble privately and conduct experiments. In England meetings were held in Oxford and London, and, upon the restoration of the monarchy in 1660, the new king, Charles II, was asked to sponsor a scientific society. The king's indifference to pure science, which has been attributed to "his limited mind," was in fact based on an acute assessment of the interests of the public and the government and an accurate judgment that most of the work that the organizers of the projected society had been doing would prove to be socially useless. When the celebrated pneumatic experiments that led to Boyle's Law and to a clarification of the role of air in respiration and combustion were demonstrated to him, Charles scornfully dismissed them as the mere "weighing of ayre" and urged the experimenters to redirect their efforts in order to extract useful results from their discoveries (Merton, 1970: 166). Boyle's Law was a profound abstraction, which, however, was to remain essentially inapplicable to industry, agriculture, or medicine for nearly two centuries. And the ponderously demonstrated fact that organisms and fire both require air must have confirmed the king's impression that the new science was indeed an idle amusement. Boyle's place in the history of seventeenth-century science is secure, but not in the histories of technology and industry prior to the middle of the nineteenth century.

Charles II's own interest in nature was in a decidedly utilitarian direction. He was "most graciously pleased" by investigations aimed at finding longitude at sea, and he took a personal interest in occult research: "King Charles II, who also had his own laboratory, was as fascinated by astrology as by astronomy. Research in chemistry might come out of the age-old pursuit of alchemy, the elusive art of manufacturing precious out of base metals" (Ashley, 1961: 155). In 1662 the king was wheedled into granting a charter establishing the Royal Society of London for the Promotion of Natural Knowledge. He had been assured that he would thus become the "Founder of some thing that may improve practical and Experimental knowledg [sic], beyond all that has been hitherto attempted, for the Augmentation of Science, and universal good of Man-kind."[20] Evidently, however, Charles remained skeptical, for he granted the charter, but nothing more. Even with its charter, the Royal Society remained, during the seventeenth century and in large measure in the eighteenth as well, essentially a club, accommodating the interests of its members, rather than an institution, serving the interests of the public.[21]

Although the abstract purity of much of the new science of the sixteenth and seventeenth centuries held no appeal for the governments of Europe, and its presumptive utility remained unconfirmed, its intellectual triumphs were undeniable; and these were what presumably raised the status of the savants in the eyes of government authorities. Heliocentric astronomy, Kepler's laws, the discovery of the circulation of the blood, the law of falling bodies, the calculus and analytic geometry, the inverse square law of universal gravitation, Hooke's and Boyle's Laws— these were surely impressive, especially coming as they did in rapid succession. Might they not somehow be useful? The principle that knowledge is power was commonly accepted, and had for millennia authenticated the claims of both rational and occult research. Now it would need to be confirmed and demonstrated for the novelties of the new science. The possibility of harnessing the pure knowledge of this new breed of savants to the objectives of social betterment seemed clear to some (although not, for the most part, to the researchers themselves), but the actualization of that possibility was to prove to be extraordinarily elusive.

Having reached this historical moment, when modern science, in the era of Isaac Newton, was acquiring its definitive shape, it is worth lingering on the difficulty of reconciling the scientist's love of pure research with the public's insistence upon useful knowledge. That passion for unfettered inquiry, affirmed by the title of Newton's great book *Mathematical Principles of Natural Philosophy*, was confronted at every turn by the state's preference for investigations that would result in palpable benefits. As the governments of Western Europe became increasingly centralized, they asserted their demand that science contribute to social betterment in economic, political, medical, or military terms in return for any institutional support it might receive. But the scientists of Europe had inherited a rich Hellenic tradition and had already nurtured it for three hundred years, under the indulgent auspices of the university and private patrons. Many of them, including most of the giants of the Scientific Revolution, were immersed in that tradition of pure research driven mainly by intellectual zeal. Those two incongruous traditions, the system of state-sponsored institutions that was pioneered in the ancient and medieval East and the individual scientist's Hellenic passion to know the natural world, inevitably came into sharp conflict.

At the founding of the Royal Society, the writings of Francis Bacon had been "the great formative influence on the Society's concept of science" (Purver, 1967: 6). Its charter members, at least publicly, were committed to Bacon's two principal canons of scientific research: the method of induction, whereby theories, without any preconceptions, would "arise" from cumulative experiments and observations; and his insistence on the priority of applied research—a combination of "experiments of light" and "experiments of fruit." On both counts the society would fail; and its failure would lead, not to a repudiation of the new learning, but rather to the direct intervention of government in the procurement of scientific services.

Less than twenty years after the society was chartered, Robert Hooke (1635–1703), who was unambiguously devoted to Bacon's antimetaphysical inductive method, unambiguously violated its fundamental premises by proposing a deeply metaphysical theory of elasticity, beyond the reach of any possible experiments. It postulated that all bodies are immersed in "a fluid subtil matter" that forms "the Menstruum in which they swim" (Hooke, 1678: 341–42). Prior to the founding of the Royal Society, Hooke had already alluded to his theory of an "ambient fluid," and had invoked it to explain gravity as well as elasticity—and, indeed, almost everything else: "Nay," he wrote, "I know not whether there may be many things done in Nature, in which this [Ambient fluid] may not (be said to) have a Finger" (Hooke, 1661: 41). In the face of this glaring inconsistency between his theory of the ambient fluid and Bacon's condemnation of philosophical speculation and theory-mongering, he quibbled:

In the meantime, I would not be thought guilty of that Errour, which the thrice Noble and Learned Verulam [i.e., Francis Bacon] justly takes notice of. . . . For I neither conclude from one single experiment, nor are the Experiments I make use of, all made upon one Subject: Nor wrest I any Experiment to make it *quadrare* with any preconceived Notion. (Hooke, 1661: 41)

Around the time that Hooke was enmeshed in these metaphysical obscurities William Petty (1623–1687), another charter member of the Royal Society, who also endorsed Bacon's repudiation of speculative theories, was conjecturing that elasticity is the consequence of the coitus of male and female magnetic atoms—"even without a Metaphor"! (Petty, 1674: 130–32).[22] The shade of Bacon must have winced. Where Hooke had pretended that his notion of an ambient fluid was somehow derived from experiment and denied any violation of Baconian principles, Petty admitted that his transgression was a departure from Baconian norms: "For I have in this Exercise declined all Speculations not tending to practice, and ventured at few Hypotheses, but that of *Elasticity*" (Petty, 1674: 5).

But Bacon's inductive method was perhaps most seriously discredited by the triumph of the Newtonian notion of physical "attraction" across empty space, despite Newton's lame Baconian disclaimer that he formulates no abstract theories. After the first ten years of the Royal Society's existence the prescriptions of the inductive method were regularly transgressed by scientists who were passionately driven by an abstract fascination with the processes of nature. Inductivism remained a programmatic ideology, but it was flagrantly neglected as a method of research.

The Baconian program of applied science fared no better than the inductive method. Seventeenth-century England was in the grip of a timber famine, brought on mainly by the excessive cutting of what was in the British Isles only a modest resource to begin with (Clow & Clow, 1956). Early in the century, the shortage had become so critical that James I issued a *Proclamation Touching Glass*, prohibiting the use of wood as fuel in the making of glass and urging the population to return to the "ancient manner of drinking from stone cups." By mid-century the crisis had intensified (partly because glassmakers flouted the king's prohibition), and it became necessary to import timber from Norway, Poland, Prussia, Danzig, and Bohemia.[23] The Royal Navy, concerned about the supply of its most vital material and laboring under the illusion that the scientists at the newly chartered Royal Society could do something about it, challenged the society's members to find a solution. Here was an opportunity to transform the society from a club into an institution, and to do so by following Bacon's injunction that "the true and lawful goal of science is the endowing of human life with new discoveries and powers." The society responded by inducing one of its charter members, John Evelyn (1620–1706), to conduct a pioneering study of sylvaculture (Avignon, 1971: 495) and by initiating a systematic experimental program to study the strength of structural timber. The utter inability of seventeenth-century science to come to the aid of industry was soon apparent. The prescriptions of Evelyn's *Sylva* could do little to reforest the British Isles, and the society's experiments on the breaking of wood, following the Baconian inductive method, reached a fundamentally incorrect formula for the strength of timber beams. Even worse, Galileo had already derived an essentially (but not strictly) correct result through the application of theoretical statics without conducting any tests or experiments at all (Dorn, 1970: 72–84). In the end, builders and engineers, with full confidence in their rules-of-thumb, derived from a long tradition of craft experience, greeted both results, the incorrect and the correct, with equal indifference.

By the eighteenth century, within a few decades of its founding, it was clear that applied science was not the Royal Society's métier. As an institution obligated to perform public services it had largely failed, since no way could be found of effectively focusing its results and achievements on social objectives. Baconian rhetoric and royal exhortation were evidently not enough. Henceforth, the hope of finding solutions to practical problems called into play direct government action.[24]

A problem that had been distressing European mariners since the beginning of long-range east-west voyages was the inability to determine longitude at sea. In the Northern Hemisphere, the determination of latitude is a relatively simple matter, the latitude being equal to the altitude of the North Star. But without the ability to track time aboard ship, a determination of longitude could not be made, and crossing the Atlantic was therefore an uncertain operation. When some of the most eminent scientists of Europe, including Galileo and Christian Huyghens, applied themselves and their astronomical charts to the problem without success, the governments of the chief maritime nations of the Atlantic seaboard took the lead and offered substantial rewards for a solution.

Early in the eighteenth century, the British government announced a reward of £20,000 and established the Board of Longitude, a governmental institution, to supervise the search for a method of determining longitude that would be workable on a rolling and pitching ship under wildly variable atmospheric conditions. In the course of the century many solutions were proposed, until the problem was finally solved—not by a scientist but by a clockmaker, John Harrison (1693–1776), who devised a clock that employed counteracting balances to negate the effects of a ship's motion and a thermocouple to compensate for temperature changes. The success of the clockmaker along with the failure of the savants could hardly be surpassed as a revealing demonstration of the superiority of eighteenth-century craft knowledge over theoretical knowledge in the solution of practical problems.[25]

Throughout Europe, although the pattern varied, the same failure to apply theoretical science to economic betterment prevailed. During the eighteenth century, scientific organizations proliferated, frequently as the result of attempts by governments to find the right combination to extract public benefits from the knowledge and expertise of the learned members. Many of the newly founded organizations were in fact explicitly utilitarian, dedicated to the application of knowledge to agriculture, medicine, or technology (McClellan, 1985: 38–39). Botanical gardens were established and botanical expeditions were financed; observatories were built and astronomical surveys were organized; and prizes were offered and awarded for the solution of scientific and technical problems. The old institutional structure, centered in the medieval university, had proven inadequate for the social utilization of science. In its place a new structure was built based on learned societies supported in part by the state (McClellan, 1985: xx). Through the new institutions the state undoubtedly derived benefits in the form of *services* supplied by the community of scientists; but the application of *research* to the solution of technical problems, along the lines that Bacon had called for, remained a program that would be realized only in the nineteenth and twentieth centuries.

Although the range of services that science could offer society was still narrow, a trickle of uncertain expenditures flowed from the state to scientific institutions, mainly, as might be expected, in support of useful activities. In sixteenth-century Spain, under Philip II (r. 1556–1598), the pattern was already clear. The Spanish monarchy was the most centralized government in Europe, and it acted with firm authority in demanding that knowledge be applied to national needs. Artillery schools were in existence, and in 1582 the crown founded and financed the Academy of Mathematics. Despite its name, however, the curriculum was centered around navigation and military engineering, while mathematics instruction was on the most elementary level. Euclidean propositions were presented to provide the basis of the "theory of proportions" (which, in the sixteenth century, amounted to geometrical rules-of-thumb) and to lay out the polygonal fortresses that gunpowder artillery had made necessary; and arithmetic was taught "for calculating the costs of construction and many other things." Artillery instruction was rudimentary and the theoretical pretensions of some of the instructors were inconsistent with the realities of contemporary gunnery (Goodman, 1988: 125–29).[26]

The curriculum also included occult studies, since Philip was moderately interested in astrology and firmly devoted to alchemy. Both attitudes were shaped by utilitarian criteria rather than philosophical interest. The claims of astrologers were so frequently discredited that Philip maintained "a middle position of restrained curiosity." In alchemy he was skeptical about the claims of metallurgical alchemists, whose attempts to transform base into precious metals had consistently proven futile, but he remained committed to what he took to be the successes of medical alchemy, for, then as now, more patients survived than succumbed regardless of the treatment (Goodman, 1988: 3–14).

The Spanish crown also patronized hospitals and cosmographers. This support, too, was typically skewed towards public welfare and technology. But the commitments were invariably limited by the state's unwillingness and inability to allocate large sums of money to the effort: "Shortage of money has been shown to be a constantly recurring theme in all matters relating to crown technological projects. . . . The failure of the treasury, drained by inherited debts and expanding warfare, was the most important reason for Spain's limited technological achievement" (Goodman, 1988: 262–63).

For seventeenth-century France a similar pattern is evident. Cartography was a field that battened on bureaucratic interest, as geodetic surveys, in the service of military engineering and economic development, were sponsored by the state. At the end of the century the highest-paid member of the newly founded Académie Royale des Sciences was Gian Domenico Cassini (Cassini I: 1625–1712), an astronomer and cartographer. He assured the king (Louis XIV), "a patron who [like Charles II] had no head for scientific theories, but who wanted better maps, better health, and better military equipment," that his research would promote utilitarian progress in navigation and mapmaking (Stroup, 1987: 53). In contrast with the English government, which had declined to fund the Royal Society, the French government subsidized the Paris Académie from its founding, in the 1660s; but it did so on a modest level and with a fluctuating degree of commitment. In the 1680s Cassini I carried out geodetic surveys as part of a large-scale project to map the entire kingdom, but funding was soon discontinued and the work was abandoned (Taton, 1971a: 103; Stroup, 1987: 54). Two generations later, under the direction of Cassini I's grandson, César-François Cassini de Thury (Cassini III: 1714–1784), the project was revived, at a projected cost of 700,000 livres, but again the government withdrew financial support and interrupted the work (Taton, 1971b: 108).

An indication of how little the French government was willing to spend on science at a time when its demonstrable ability to serve the state was slight is that during the last third of the seventeenth century the average annual expenditure on the Paris Academy, including all stipends, support of the observatory, and funding of research projects, was 63,000 livres, only seven times the salary paid to Cassini I (Stroup, 1987: 152). And in return for these modest grants the government, as always, expected social or economic benefits, and directed the efforts of the academy towards the acquisition of useful knowledge and the rendering of useful services. In concluding her detailed analysis of royal funding of the Paris Academy during its first few decades, Alice Stroup stated, "Two features of the Academy, its usefulness and its relative cheapness, helped guarantee its survival" (Stroup, 1987: 51).

On the strength of the historical records for England, Spain, and France it is safe to assume that the governments of Europe were motivated to support science, insofar as they did so, because of its low cost and its hoped-for utility. Although that utility, from the sixteenth through the eighteenth centuries, may have been demanded by the state and sometimes intended by the savants, it was in fact achieved only along a few lines of endeavor—agricultural and botanical research, cartographic and navigational improvements, and perhaps some tangential influence on the design of fortifications, on ordnance, and on ballistics. But the work of the giants of European science remained barren as a source of practical benefits. In Europe, scientific research was still Hellenophilic and individualistic; it was only the network of institutions—the universities and the learned societies and academies—that was Asiatic and that occasionally supplied technical services and minor improvements. The major technical innovations and improvements were derived, not from science at all, but rather from the ingenious tinkering of craftsmen and engineers.

Although the new science was highly abstract and displayed no concrete ability to secure practical benefits, it evidently engendered hopes that it would provide them in the future or, at least, that its practitioners were competent to provide useful services. It has remained a principle of conventional wisdom, albeit unconfirmed by the historical record, that science is intrinsically technology's tutor, and that progress in the crafts and technique is consistently furthered by scientific theory. That the Industrial Revolution of the eighteenth century followed hard on the heels of the impressive scientific developments of the seventeenth century has reinforced the persistent misunderstanding that conflates science and technology and postulates that they stand in the relationship of benefactor and beneficiary.[27]

If the process of industrialization in the eighteenth century were as dependent on the application of theoretical science as this misunderstanding postulates, the monarchs of Europe would have been irrational and dysfunctional in stinting on their patronage of institutions that sponsored scientific research. (During the eighteenth century and well into the nineteenth the British government continued to deny any financial support to the Royal Society.) In fact, however, the technological history of the Industrial Revolution shows that the technical innovations owed little if anything to scientific principles. Much of the scientific theory that might have been applicable to the development of those technical novelties was indeed not formulated until more than a hundred years later. The steam engine, invented by Newcomen and improved by James Watt, was the result of inspired tinkering. Watt was a member of the Royal Society and published a few articles in its *Philosophical Transactions*, but none on any theory of the steam engine. Indeed, none of Watt's inventions—the separate condenser, the double-acting engine, the parallel motion, sun-and-planet gearing, steam cutoff, or the indicator card—was based on theoretical principles; they were all the products of his remarkable power of originality as an inventor.[28]

Thermodynamics and kinematics, the two sciences that in retrospect might appear to have been applicable to steam engineering, were developed only in the nineteenth century, more than a hundred years after the invention of the steam engine. In 1824, 112 years after Newcomen's invention, Sadi Carnot (1796–1832), a founder of thermodynamics, lamented in his seminal work, *Reflections on the Motive Power of Fire*, that, although the importance of steam engines is "enormous, their use is continually increasing, and they seem destined to produce a great revolution in the civilized world," their "theory is very little understood, and the attempts to improve them are still directed almost by chance" (Carnot, 1824: 3, 5). Instead of principle inspiring practice, the reverse was true—the steam engineering of the eighteenth century led theoretical scientists to conduct research and originate the sciences of thermodynamics and kinematics in the nineteenth century.[29]

Iron metallurgy, which was revolutionized in the eighteenth century by the invention of the process of smelting iron ore with coal instead of wood and which contributed significantly to the Industrial Revolution, owed as little to theoretical science as did steam engineering. The improvements were made by ironmasters steeped in the traditional methods of the crafts. James Conant, in a study of eighteenth-century chemistry, concluded that

the advance in science and the progress in the practical arts are both rapid; yet the two borrow relatively little from each other in the way of concepts or new factual information. . . . The application of science to industry . . . still lies in the future. . . . It is important to remember that the whole revolution of the making of iron and steel in the eighteenth century was based on purely empirical experimentation. The substitution of coke for charcoal, the invention of the crucible steel process, the improvements in iron-ore smelting that yielded pig iron, the development of the puddling process for making malleable iron (mild steel) were all accomplished without benefit of science. (Conant, 1957a: 112)

And the improvements in the textile industry, the innovative spinning and weaving machinery that is traditionally associated with the Industrial Revolution in Britain, were equally innocent of scientific theory. (Indeed, what contemporary theory could have favored such technical developments?)

An event much more revealing of science (or, rather, its inapplicability to industry) in the Industrial Revolution occurred in 1800. At the end of the eighteenth century there were still no scientific institutions in Britain (or anywhere else) that could confidently be expected to supply analytical solutions of engineering problems. With the Industrial Revolution in full swing, the Port of London proposed to build another bridge across the Thames, to serve the growing metropolis. One of the designs submitted was a spectacular and unprecedented plan by Thomas Telford (1757–1834) to construct a cast-iron bridge in the form of a single arch 600 feet in span. In 1800 metal bridges were still a novelty, and there were no traditional methods or rules-of-thumb that could guide the design. There was also no institutionalized applied science to which the problem could be submitted. Instead, Parliament attempted to fill the institutional void with an ad hoc institution in the form of a select committee divided into "theoreticians" and "practitioners." It was hoped that applied science would appear as a synergism arising from the separate deliberations of theoretical scientists and experienced, but unschooled, craftsmen and engineers. The results were sobering. They confirmed the difficulty of institutionalizing theoretical science and bringing it to bear on the solution of practical problems. The "theoreticians" lacked even a clue for analyzing the stresses in a built-up fixed arch. The Astronomer Royal recommended that "the Bridge be painted white, as it will thereby be least affected by the Rays of the Sun"; and the Savilian Professor of Geometry at Oxford calculated the length of the arch to ten millionths of an inch and its weight to thousandths of an ounce. The "practitioners" (who included the engineers James Watt and John Rennie), while more sensible, were unable to provide analytical grounds for their opinions, and could only offer a few suggestions about how to fabricate the structure.

Science had made spectacular gains during the previous centuries, but despite sporadic efforts, society and its industries were still unable to bring those lofty achievements down to earth. In the end, war intervened and the bridge was never built. In its place was left the distinct impression that practical problems still required practical solutions found by individuals schooled in experience and intuition, and not in theoretical science. John Playfair (1748–1819), who was professor of mathematics at the University of Edinburgh and a member of the committee of "theoreticians," conceded in his report that mathematics and analytical mechanics remained inapplicable to the problems of the engineer. "It is therefore from Men bred in the School of daily Practice and Experience," he concluded, "and who, to a Knowledge of general Principles, have added, from the Habits of their Profession, a certain *feeling* of the Justness or Insufficiency of any Mechanical Contrivance, that the soundest Opinion on a Matter of this Kind is to be obtained" (British Sessional Papers, 1801: 25; this source contains the reports of the members of both committees). Against this background, the monarchs of eighteenth-century Europe can be seen to have been acting rationally when they patronized science with a tight fist.

But the belief that innovations in technology, engineering, and industry are characteristically derived from theoretical science persists and occasionally surfaces in the writings of cultural historians. In its most recent expression, a colorful account of the Scientific Revolution and its consequences in the eighteenth century, the idea that science played a significant part in the Industrial Revolution is reformulated in phraseology about the "cultural origins" of industrialization: "The science of the English philosophes appears not only as a unique version of Enlightenment but also as the historical link between the Scientific Revolution in its final, English phase and the cultural origins of the Industrial Revolution" (Jacob, 1988: 139–40). We are warned that "to imagine that the scientific learning . . . had little direct relevance to that historical development we call the Industrial Revolution is to ignore the rich historical evidence now available for the application of science to commerce and industry." Historical explanations of industrialization that emphasize the role of material conditions and craft traditions while minimizing the alleged impact of intellectual history are rejected for their "simplicity and monocausality. . . . They are not only flawed, they are also frequently boring and ultimately irrelevant" (Jacob, 1988: 140–41, 220). Instead, it is maintained that the craftsmen and engineers of the Industrial Revolution "thought their way to industrialization" in a process that, but for a conspiracy of historians, would properly be seen as part of intellectual history—an "intellectual history [that] has largely gone unwritten because economic and social history, and modernization studies, have dominated the study of the first Industrial Revolution" (Jacob, 1988: 136).

To illustrate "the rich historical evidence now available for the application of science to commerce and industry" in the Industrial Revolution, we are offered an account of the improvement of the harbor at Bristol. John Smeaton and his disciple, William Jessop, a minor figure in the history of British engineering, who are described as "probably the best engineers of their day," were consulted on the project. (Smeaton was indeed an outstanding engineer of the period, but Jessop, a contemporary of James Watt, John Rennie, and Thomas Telford, was a long way from the first rank.) The only reference to a scientific principle in any of the recommendations was a single calculation allegedly based on the Galilean law of falling bodies, which, according to Jessop, "has been found by experiment." In testifying about his exploit, Jessop conceded that he had forgotten "the Theory" (he offered to recall and supply it "in a few Weeks" [Jessop's italics]) and had "contented [himself] with referring to certain practical rules . . . deduced therefrom, and corrected by experience and observation." Moreover, he commented that the action of falling water deviates from the rule and "only experience and nice observation can nearly ascertain" the degree of the deviation (Jacob, 1988: 220, 226–33). If this application of "theoretical science" to engineering typifies "science in the Industrial Revolution," then the intellectual history of that revolution has been properly neglected.

The eventual improvement of the harbor at Bristol, which was carried out early in the nineteenth century, was based on a recommendation Smeaton had made—"enclosing a long bend of the river between lock-gates and cutting a by-pass for the river Avon" (Hamilton, 1958: 468). Such marine works, which followed long-established practices, were performed without any application of theoretical hydrodynamics or any other science. During the second half of the eighteenth and first half of the nineteenth centuries, hydrodynamic theory and hydraulic practice remained separate enterprises and, because of the eighteenth-century development of abstract mathematical hydrodynamics, may even have been diverging: "Hydrodynamics and hydraulics thus went their separate courses . . . the one becoming an ever more elegant mathematical discipline and the other an ever more useful engineering art" (Rouse & Ince, 1957: 112). The attempt to honor the craftsmen and engineers of the eighteenth century by linking their vocations to theoretical science in fact defames them, by implying that their craft and technique were ineffective without theoretical enlightenment.[30]

The European Industrial Revolution is an episode in social, technological, and economic, but not intellectual, history. The new science wrought by the mathematicians and natural philosophers of Europe was a sparkling ornament that enlightened its educated citizens and encouraged its rulers to intensify the attempt to institutionalize science in the hope of deriving significant social benefits, but the philosophers of nature were not yet ready to integrate "experiments of light" and "experiments of fruit." In the meantime, the machines and mills, the bridges and canals, and the river and harbor works of the Industrial Revolution were wrought by another class of men. It was only during the second half of the nineteenth century that European society steadily exacted more than honor, *gloire*, and illumination from theoretical science. Only then did science and technique converge—electrodynamics and the production of power and light, thermodynamics and heat engines, kinematics and machine design, hydrodynamics and hydraulics, chemistry and the dye industry, the germ theory of disease and medicine—theoretical developments sometimes leading technical developments and sometimes the reverse. And the convergence was seen as a validation of faith in the social utility of science.

Only over the past 150 years did the collaboration of science and the state begin to reap the "experiments of fruit" that Bacon had envisioned. Only then did European science begin to fulfill its contract with the modern nation-state. But even then the harvest of practical benefits was at first meager, and although the state began to supply increased financial support it still did so sparingly. In Britain the level of funds that flowed from government for the support of research remained incredibly low throughout the nineteenth and into the twentieth centuries. The royal scientific societies received annual stipends of only hundreds of pounds. And even the Royal Society of London, which took the lion's share of public money, received annual grants of less than £5,000, until 1920 (Alter, 1987: 19–21). Insofar as the state in nineteenth-century Britain was willing to provide even these trifling stipends, it was almost exclusively interested in short-term results that had "obvious economic or military value." Accordingly, basic science came in for more than its share of neglect.[31] It was only after World War II that the interventionist state began to lavish support on scientific research. (The British government did not create a ministry of science until 1959.)

On the relationship between science and government, the historian Charles Gillispie observed:

From science, all the statesmen and politicians want are instrumentalities, powers but not power; weapons, techniques, information, communications, and so on. As for scientists what have they wanted of governments? They expressly have not wished to be politicized. They have wanted support, in the obvious form of funds, but also in the shape of institutionalization and in the provision of authority for the legitimation of their community in its existence and in its activities, or in other words for its professional status. (Gillispie, 1980: 349)

These "powers" that are now credited to the rational sciences have finally and fully disgraced what have come to be termed the pseudosciences, the occult learning that for four or five thousand years provided much of the justification for the patronage of research.

Because of Europe's unique environment, the centralization of authority that finally sponsored European science developed along lines radically deviant from those of governmental patronage in the ancient and medieval East. European science was patronized and institutionalized first by universities and the Roman Catholic Church and then by nation-states centralized by a variety of economic activities and governmental responses to technical developments and military imperatives—by enormously expensive gunpowder armies and the defensive systems they engendered, by complex and costly men-of-war, by timber famine so severe that it required governmental intervention, by rapid industrialization, and by overseas empire.

Prior to the intervention of the nation-state, the science of modern Europe, developing in a region of rainfall agriculture where no central authority was needed to manage the basic economy, resembled the Hellenic pattern and displayed a bias towards abstract Aristotelian research. Because of Europe's geography, the centralizing tendency of hydraulicized agriculture was evaded for long—but not forever. Finally, in a recapitulation of the Asiatic economies, the cycle was completed as the hydraulicization of agriculture was added to the repertory of centralizing forces acting on modern industrial societies. Even the most productive agricultural systems in the world, on the most fertile soils, could not sustain large and growing populations on the strength of dryland farming methods alone. Between 1800 and 1970, irrigated acreage in the world increased twenty-five times, and by the end of that period nearly 20 percent of the total was in Europe and North America (Framji & Mahajan, 1969: cxiii–cxviii). It is, indeed, only through the hydraulic intensification of its agriculture on an industrial scale that the West has been able to achieve and maintain its spectacular prosperity.

It should not be thought, however, that the hydraulic hypothesis can apply to European populations in only this limited and belated way. Where the physical environment warrants it, that hypothesis can be as descriptive of the social and scientific development of a European society as of any Eastern civilization. To observe the full process in miniature, to observe a European population as it developed a hydraulic society and cultivated a scientific tradition under physical conditions more reminiscent of the ancient Oriental kingdoms, it is necessary to follow science on its next great diffusion—westward across the Atlantic.

CHAPTER SIX

WESTERN SCIENCE GOES WEST

SAINTS AND SCIENTISTS

The United States is divided into two physical zones by the ninety-eighth meridian, which passes through Oklahoma City and through San Antonio, Texas. East of the meridian, the relief is low, there is extensive prairie and woodland, and rainfall is adequate to sustain dryland farming. Agriculturally, it is an extension of Europe, and it presented the European settlers with a familiar ecological landscape. West of the meridian, the terrain rises, and humid conditions give way to desert—the Great Salt Lake Desert, the Painted Desert, Mohave Desert, Harqua Hala Desert, Black Rock Desert, and Death Valley. Over great stretches of the American West, mean annual rainfall is less than ten inches, and over most of it less than twenty inches. Except for the coastal region of the Far West, the territory, if it is able to be farmed at all, is suitable only for the hydraulic agriculture that is made possible by its resources of surface and subsurface water. Physically, the New World, with arid lands on its western side, is the mirror image of the Old, where it is the East that is dry. In the New World, East is West and West is East.

In accordance with their ecological condition the eastern territories of the United States became the land where adequate rainfall relieved the central government of any need to manage the agricultural economy, where the independent family farm flourished, and where Jeffersonian democracy took hold. Conversely, the arid West, laced with rivers and streams and perched on aquifers, has from the beginning turned to hydraulic agriculture for its development, and has thereby led the nation in embracing bureaucratization, centralization, federalization, and agribusiness. Psychologically and culturally the European settlers of the American West initially shared the same customs and attitudes as their eastern compatriots, but the restrictions and the opportunities of material conditions imposed a sharply divergent way of life. To harness its rivers and claim its bounty, the new West adopted the traditions of the ancient East. The taming of rivers was still beyond the capacities of a decentralized market economy. It required the heavy hand of state intervention, and western settlers were not long in calling for it:

The Columbia, the Missouri, the Rio Grande, and the Colorado all eluded their grasp. So they raised their voices in one loud, sustained chant that could be heard all the way to Washington, D.C.: "We need the State!" And the federal government responded by passing the National Reclamation Act in 1902. It has been the most important single piece of legislation in the history of the West, overshadowing even the Homestead Act in the consequences it has had for the region's life. The West, more than any other American region, was built by state power, state expertise, state technology, and state bureaucracy. (Worster, 1985: 130–31)

The Bureau of Reclamation has commanded more power and exercised more authority than even a dynasty of Oriental despots. By the 1980s the seventeen western states encompassed one-tenth of the world's irrigated acreage, with more than fifty million acres, and consumed more than 90 percent of all irrigation water in the United States (Engelbert, 1984: 5).[1] After three-quarters of a century of operation, the Bureau of Reclamation could boast that it had constructed 322 storage reservoirs, 345 diversion dams, nearly 15,000 miles of canals and 35,000 miles of laterals, 174 pumping stations, and "49 power plants marketing more than 50 billion kilowatt-hours a year over 16,240 miles of transmission lines. . . . For scale of engineering, for wealth produced, the American West had become by the 1980s the greatest hydraulic society ever built in history. It had far eclipsed not only its modern rivals but also its ancient ones, Mesopotamia, Egypt, Mohenjo-Daro, China, and the rest" (Worster, 1985: 277, 276).

There can hardly be a more compelling confirmation of the power of geography in shaping society. The American West imposed its hydraulic rule on every agricultural society that took up residence on its desertified territories. The Hohokam of Snaketown and the Anasazi of Chaco Canyon, immigrants from hydraulic societies to the south, adapted readily to the requirements of irrigation agriculture. A more telling case is that of the earliest European migrants from the more humid districts of the eastern United States who had to adapt to the ways of dry-zone farming before they could settle down in the Great American Desert. Lacking most of the traditions and experience called for by agriculture in arid regions, they crossed the ninety-eighth meridian and were ineluctably pressed into a hydraulic mold by physical conditions that reshaped custom and culture. The first of these European settlers to successfully create a hydraulic society in the American West, including the monumental architecture and the political and scientific institutions that have characteristically been part of such societies, were the Latter-day Saints.

Mormonism was founded in 1830 when the followers of Joseph Smith (1805–1844) established the Church of Jesus Christ of Latter-day Saints, in an area of New York State that was the scene of intense religious ferment. Smith was a young farmer with little education who reported divine revelations and is credited by his followers with translating the Book of Mormon from golden plates to which he was directed by the angel Moroni. Mormonism repudiated both Protestantism and Catholicism, and from its beginnings it embodied communitarian ideals and practices, a unity of spiritual and temporal life, and an emphasis on economic matters, all of which would later facilitate its adaptation to the demands of cultivating a desert.

Over the next twenty years the Saints were buffeted by rejection, persecution, envy, hostility, schism, and violence, and were driven steadily westward from one attempt to build an American Zion to the next, from New York to Ohio, Missouri, Illinois, and finally to the desolate safety of the Great Basin of the West, a territory over 200,000 square miles in extent, 95 percent of which is mountain or desert. Although the land was then a province of Mexico, it was for the most part unoccupied, and that condition accounted in large measure for its appeal to the immigrants, just as absence of competition for stricken lands had, over the millennia, served as a beacon for other peoples who became irrigation farmers in arid regions: "A large area over which there would be little dispute or difficulty was the Great Basin. Nobody would envy the Mormons their possession of it" (Arrington, 1958: 40).

In 1844 Joseph Smith was lynched in Illinois (after announcing his candidacy for the presidency of the United States), and leadership of the Mormons passed to Brigham Young (1801–1877), a former painter and glazier, who in 1847 led the main body of Saints into the Salt Lake Valley on the last leg of their troubled wandering.[2] It was a land designed for the practice of hydraulic agriculture by any people who could devise an appropriate social organization. It received less than 15 inches of rainfall annually but was blessed with both a profuse network of streams and stretches of potentially fertile alluvial soil. The only component of hydraulic agriculture that was lacking was one that the Saints were uniquely prepared to provide, communal effort governed by a central authority. On their great odyssey from New York to Utah, the Saints had experimented with collectivist social and economic policies— communal ownership of property, redistribution of wealth to assist the poor and the newly arriving converts and settlers from the East, the organization of public works, and the centralization of authority in a church that was simultaneously a government. Those policies and practices were to be tested in their struggle to transform a desert into fertile fields.

Mormon experiments with communitarian social arrangements, even before they created a hydraulic society in the western desert, indicate how culture and ecology may interact; and they suggest that the process is not the one-way cause-and-effect relationship that the hydraulic hypothesis is sometimes accused of postulating. Conformist Christians, unlike the Saints, would not have been driven west to begin with. And religious deviationists of another stripe, lacking the Mormon communitarian experience, would not have settled in a desert, and would surely not have survived in one. It was the consistency between principles and physical conditions that directed the Mormons to their destiny.[3]

Under the direction of Brigham Young and a "Quorum of the Twelve Apostles," a town was surveyed by an officially appointed surveyor; houses, roads, and bridges were built with "public labor"; an irrigation canal was dug; "dams and ditches were constructed on a community basis"; and a "watermaster," a public authority, was appointed to allocate the vital resource on collectivist principles. By the end of their second decade in the desert the Saints had "constructed 277 canals, totaling 1,043 miles and bringing water to 115,000 acres" (Kirby et al., 1956: 457). Although land was distributed in family smallholdings, it was held on the basis of a stewardship system that embraced the principle "equal according to circumstances, wants and needs." Water was retained under public ownership. Brigham Young declared: "There shall be no private ownership of the streams that come out of the canyons, nor the timber that grows on the hills. These belong to the people: all the people." As inevitably as in the ancient Oriental kingdoms, irrigation agriculture called into being a labor corvée. When families required irrigation water, "the bishop arranged for a survey and organized the men into a construction crew. Each man was required to furnish labor in proportion to the amount of land he had to water. . . . The labor necessary to keep the canal in good repair was handled in the same way, in accordance with assignments made in regular Sunday services or priesthood meetings" (Arrington, 1958: 51–53).

The harmony between the demands of desert agriculture and the communitarian principles that came increasingly to underlie Mormon social life was clearly recognized by Great Basin pioneers, although it was stated in terms more akin to Christian communism than to Oriental despotism. John Widtsoe (1872–1952), a church leader and an eminent soil scientist, stated:

If the canal breaks or water is misused, the danger is for all. In the distribution of the water in the hot summer months when the flow is small and the need great, the neighbor and his rights loom large, and men must gird themselves with the golden rule. The intensive culture, which must prevail under irrigation, makes possible close settlements, often with the village as a center. (Quoted in Arrington & Bitton, 1979: 318)

There are indications that the desert imposed its will on institutions, ideology, and general economic arrangements as well as on agriculture and social relations. The digging of irrigation canals became a water ritual among the Utah Mormons: "the construction of water ditches was as much a part of the Mormon religion as water baptism" (Arrington, 1958: 26). The office of "watermaster" was an institution that could not have developed east of the ninety-eighth meridian: "Each bishop was appointed to be watermaster for his ward to assure ample channels and equitable division of the water" (Arrington & Bitton, 1979: 115). And timber, which was a vital material and was used for temple building as well as in economic activities, was placed under the control of the central authority: "No person was allowed to build with logs without permission" (Arrington, 1958: 48).

After heroic efforts and agonizing setbacks, the Saints succeeded, and they prospered in a habitat that would most assuredly have defeated individualist pioneers. The Mormons recapitulated in the American West the social and infrastructural features of the ancient East—in their governance, in their agricultural system, and in their monumental architecture, too. The Salt Lake Tabernacle, completed in 1875, "is an immense, elliptical, turtle-backed auditorium, 250 feet long, 150 feet wide, and 80 feet in height, which will seat 10,000 people. . . . It required more than a million and a half feet of lumber. The architect boasted that it was 'the largest Hall in the world unsupported by columns.'" An indication of the tabernacle's monumental effect is Aldous Huxley's description of it as the "Chartres of the desert" (Arrington, 1958: 213–15).[4] In the construction of temples, Brigham Young enunciated the principle of the corvée in ringing terms: "We shall expect from fifty to one hundred men every working day throughout the season to labor here. . . . If any person should enquire what wages are to be paid for work done on this Temple, let the answer be 'Not one dime'" (Arrington, 1958: 339–40).

Monumentality and large-scale enterprise were not confined to temple building and irrigation infrastructure. In its 150-year history the Church of the Latter-day Saints has grown from a few thousand beleaguered disciples to a rapidly increasing membership of over seven million worldwide. Mormon educational facilities, including Brigham Young University, have been estimated to be worth nearly two-thirds of a billion dollars. Heavy investments in communications properties, including presses and radio and television stations, are valued at over half a billion dollars. The Mormon Church is "the single largest ranching enterprise in the United States"; the value of its agricultural properties has been estimated at over two billion dollars, and its total assets in 1983 at nearly eight billion dollars (Heinerman & Shupe, 1985: 73–74, 119–20).

Historians have occasionally noted the parallels between the social and technical achievements of the Mormons and the hydraulic civilizations of the ancient East. William McNeill observed:

> The Mormons thus recapitulated the experience of the world's earliest civilizations. Religious authorities in the Tigris-Euphrates, Nile, and Indus Valleys created a social hierarchy of unprecedented power and developed human skills as never before, all on a basis of irrigation agriculture that required large-scale pooling of labor for its maintenance. In exactly the same way, Brigham Young raised the Mormon community far above levels otherwise attainable by pioneers in the desert frontiers of North America. His early concentration of public resources on monumental temple-building aptly symbolizes this link with his remote predecessors, whose ziggurats and pyramids still astound us. (McNeill, 1983: 68)

Ironically, the Saints had set out to create the "true Israel" and had, instead, under the pressure of physical conditions, created Babylon. Brigham Young, who "exercised supreme control in the cooperative theocracy," ruled a society based on hydraulic agriculture, and ruled with almost unlimited authority. When the Mormons established a provisional government in 1849, which they designated the "State of Deseret," with Brigham Young at its head, a society startlingly reminiscent of the Asiatic Mode of Production was recognizable in the Great Salt Lake Desert of the American West.

If the saga of the Latter-day Saints had been confined to these accomplishments it would be an impressive chapter in the history and geography of agricultural societies. But the Mormon experience also adds a chapter to the history and geography of science, for the Saints have uncannily achieved an eminence in the world of science that again recalls the attainments of the ancient East and of medieval Islam. Although their high distinction in science originated mainly around those sciences that served the needs of irrigation agriculture, it broadened as it developed, and it came eventually to embrace many fields of scientific research. In studies of the productivity of American scientists, conducted state by state in the 1930s, Utah ranked first. In a later study of the productivity indices of American science from 1920 to 1961, Utah was again first—indeed, in a class by itself. Utah's index was 52.5; the next ten states were clustered between 40.4 and 34.6 (Hardy, 1974: 499). Moreover, the performance of Utah, which depended on three schools—Brigham Young University, the University of Utah, and Utah State University—all heavily enrolled by Mormons (Brigham Young University almost exclusively), was evidently a Mormon phenomenon, not a regional pattern: "Compared to other states in its region, it is deviantly productive" (Hardy,

1974: 500). In still another study, this one of the productivity of American scientists and scholars according to undergraduate college, Jews and Mormons were singled out as special cases:

Especially noted were two homogeneous clusters of highly productive schools . . . (i) four New York City schools with heavy Jewish enrollment (City College, Brooklyn College, Queens College, and Yeshiva University); and (ii) three Utah schools of largely Mormon enrollment (Brigham Young University, Utah State University, and the University of Utah). (Hardy, 1974: 498)

This surprising sociological pattern of academic and intellectual achievement has sometimes been attributed to the high attainments to be expected from "people of the Book." But Jewish and Mormon scientists have related to "the Book" quite differently: while Jewish scholars and scientists are often lapsed in their religious observance, an extraordinarily high percentage of Mormon scholars and scientists (72%) have remained faithful and "actively affiliated" with their church (Hardy, 1974: 506). Correlations between these two groups based on religious attitudes are thus questionable. More suggestive correlations may perhaps be drawn between their scientific achievement and their institutional status. A high percentage of Jewish scientists over the past century have, in a Hellenic pattern, distinguished themselves in pure research, functioning as individualists embedded in what to them were the foreign institutions of their European diaspora (institutions to which they were first admitted only in the nineteenth century). Mormon scientists, in contrast, have displayed an Asiatic pattern, at first distinguishing themselves mainly in the applied sciences, often connected with agriculture or hydrology, that were obviously part of the institutional agendas of their own society.

The same imperative that applied to the cultivation of the soil and the conservation of resources was applied to the cultivation of knowledge, and for the same social reasons. The studies that reveal the Mormons' preeminence in science show that their excellence was at first especially marked in the agricultural sciences, including agronomy, botany, and soil studies. It was the book of nature, rather than the Book of Mormon, that led them to science. Whatever tension may have mounted between the principles of science and the Mormons' ardent beliefs in revelation and angels did not inhibit their pursuit of secular knowledge, particularly in fields that promised to serve their practical economic interests. On the contrary, church authorities encouraged scientific study, which at first could only be acquired in the outside world. Among the Mormons, science was promoted, not disarmed, by an all-pervasive state religion. In an observation that is apposite to any society with a state religion that includes a belief in miracles, Donald Worster referred to the inherent opposition of strict religious convictions joined to the study of the sciences:

The more Mormons controlled nature, the more they needed science and technology. . . . The Church required, for its own maintenance and growth, a capacity in people for unquestioning faith in angels and golden plates. However, it also depended on the expansion of hydraulic engineering, irrigation expertise, and sophisticated agricultural knowledge. At some risk, it sent its young people away to get the best training possible. (Worster, 1985: 81)

The first of its young people that the Mormons sent out to the world of secular science to acquire the knowledge required for intensive agriculture was John Widtsoe, a Norwegian immigrant and convert who became an apostle of the Church. He achieved high distinction in the study of biochemistry at Harvard (from which he graduated *summa cum laude* in 1894), earned a Ph.D. at Göttingen, and served first as president of Utah State Agricultural College and later of the University of Utah. As director of the Utah Agricultural Experiment Station, Widtsoe undertook the study of "irrigation from a scientific point of view" and produced a series of publications on the subject, including *Principles of Irrigation Practice* (1914); he is "often regarded as the founder of irrigation science"; and he studied dryland agriculture as well, particularly methods of retaining moisture in the soil in regions of unreliable, marginally adequate rainfall (Arrington & Bitton, 1979: 312–13, 382 n. 15).

One of Widtsoe's students, Franklin Harris, earned a Ph.D. at Cornell in soils and agronomy and later "developed a program of training that produced graduates who went on to dominate the fields of soil physics and agronomy." Harris became president of Brigham Young University, and "under his administration the institution attained national accreditation for its teaching and research." Another of Widtsoe's students became a professor of agronomy and then president at Kansas State College. Indeed, in 1977 it was estimated that "approximately half of the soil physicists in the English-speaking countries are Mormons or received their advanced degrees under Mormon professors." The same study found that, "out of the ten soil physicists whose work is most frequently cited by others, eight are Mormons, and fourteen of the top twenty [are Mormons]." Mormons also have dominated the Soil Science Society of America, the American Society of Agronomy, and the American Society of Agricultural Engineers (Arrington & Bitton, 1979: 315, 317).

The preeminence of the Saints in those sciences that are applicable to dry-zone agriculture resulted in extensive contacts between Utah and foreign countries, particularly and quite appropriately, with Iran, situated in the territory of the first hydraulic society. Early in the present century Persian students began attending the Utah colleges, and Mormon engineers have supervised the restoration of irrigation systems in Iran.[5] The kinship between Mormons and Persians was displayed not only in agriculture and applied science; it extended to ideology and theology as well: "Mormonism had much in common with Islam" (Arrington & Bitton, 1979: 316), including presumably a high regard for the Old Testament and the practice of polygamy.[6] One of the Mormon irrigation engineers who worked in Iran expressed the kinship in historical terms: "It gives one a thrill to be working over again the ground that was trod by the ancient prophets and to put water back into the ditches that have been dry so many long centuries" (Arrington & Bitton, 1979: 314).

The close parallels between the Mormon "Zion in the Desert" and ancient and medieval forms of the Asiatic Mode of Production provide a basis for reviewing the hydraulic hypothesis, especially since the origins and social formation of the Utah Mormons are known in much greater detail than the early histories of the ancient hydraulic societies. The claim that in Ceylon and elsewhere the hydraulic infrastructure sometimes required only modest inputs of labor, and hence no centralization of authority, is called into further question by the Mormon experience. The infrastructure created in the Great Basin was far less elaborate and monumental than the systems of the ancient Egyptians or Sinhalese. Nonetheless, the total effort embodied in the Mormon hydraulic infrastructure and social organization—the channeling of the desert, the impounding, regulation, and allocation of water, the drafting of labor battalions, the administration of public works, and the building of temples and a university—required a degree of political centralization that was far beyond anything generated by agriculture in the humid and subhumid regions east of the ninety-eighth meridian. The evolution of this cultural configuration as an enclave in the larger American society, which was at the time considerably less centralized, testifies to the centralizing effects of hydraulic agriculture even where the infrastructure is far less monumental than in the paradigm civilizations of the ancient East.

Conversely, objections to some of Wittfogel's extreme and intemperate formulations of his theory of "Oriental despotism" are strengthened by the Mormon experience. Wittfogel's determination to impale Soviet communism on the horn of despotism sometimes led him to conclude, with insufficient evidence, that centralized governance resulting from hydraulic intensification is inevitably and odiously despotic. Restrictions and coercion are indeed inevitable when individual interests are compelled to defer to the public good. Despotism, however, is too unforgiving a concept to apply to the society created by the Saints in the Great Salt Basin. Inherently, its political tendencies were indeed divergent from the main current of American society: "Theocracy, not democracy, and communalism, not capitalism, were its ideals." (Heinerman & Shupe, 1985: 3). But Mormonism was tempered, possibly by some vestiges of its libertarian New England heritage, and certainly by the restraints, including armed force, of the government in Washington. (In 1857 United States troops "invaded" Utah. Mormon militia skirmished with the U.S. Army in the Utah War, and Brigham Young was replaced as governor by force of arms.) Comparable restraints may have softened highly centralized states elsewhere, as they certainly have in the case of modern communism (after its Jacobin phase) and as they evidently did in ancient Egypt as well. Bruce Trigger observed that in Egypt "government was personal even at the highest level; access to authority was not difficult, and the right to make an appeal to Pharaoh in person was [taken for granted]. . . . On balance, one leaves the study of the subject with the firm conviction that there are, and have been, far worse systems of government" (Trigger, et al., 1983: 337).

The debate over the hydraulic hypothesis has over-simplified the issues and polarized them into caricatures of freedom and coercion. The longevity, over hundreds and even thousands of years, of various forms of an Asiatic Mode of Production, and the steadfast dedication of the Mormon pioneers under a centralized authority from which they could easily have fled (and some did), suggest that coercive controls, when they are uniformly applied and recognized to be socially purposeful, are more tolerable than the rhetoric of the "free society" has traditionally claimed.

BABYLON, U.S.A.

In the nineteenth century the centralization of authority in the United States became most intense under Mormon governance in the Great Salt Basin; but it was not confined to the West, nor was hydraulicization of agriculture its only wellspring. While the Saints were building their American Zion in the Salt Lake Valley, the eastern United States, under the pressures of industrialization, was entering its own era of federal bureaucratization.

In the 1850s, concern over steamboat explosions led to public regulation of boiler construction, the first instance of engineering design coming under federal control (Burke, 1966). Over the next century the central government in Washington intervened repeatedly in the process of industrialization; and it is ironic and significant that its intervention, involving direct ownership of vast properties and facilities, was most extensive and controversial in the modern manifestation of hydraulic technology—the production of hydroelectric power. Much of the nation's electric power is produced by transforming the energy of falling water into electricity using some of the same devices—dams, canals, and sluices—that underpinned irrigation agriculture in the Bronze Age and, at that time and since, tempted the state to play a central role in economic and social life. Because of their scale, their alteration of publicly owned river systems, and their extensive social and economic functions and consequences, large hydroelectric power stations unavoidably entail government control. Private capital has always favored relatively small projects and has been unwilling and unable to undertake the construction of regional hydraulic systems, partly because of the exceptionally large investments required and partly because of the noncommercial public functions they are expected to perform—flood control, inland navigation, and recreation—along with the production of saleable power and light. In traditional agricultural societies such large-scale public efforts have produced variations of an Asiatic Mode of Production; in modern industrial settings, even in societies that are committed to the principles of a market economy, the construction of hydroelectric power stations has resulted in government control and ownership of many facilities, including some of the largest industrial enterprises in the world. In the United States this technological imperative led, after World War I, to the federal government's construction and ownership of Wilson Dam, at Muscle Shoals, Alabama, and in the 1930s to the creation of that leviathan of centralization, the Tennessee Valley Authority.

The congressional debate over the regulation of steamboat boiler design occurred in 1852. During the debate an opposition senator asked: "What will be left of human liberty if we progress on this course much further? What will be, by and by, the difference between citizens of this far-famed Republic and the serfs of Russia?" (quoted in Burke, 1966: 21). Despite the senator's heartfelt query and the fact that federal regulatory agencies were constitutionally unwarranted, the United States embarked on its trajectory of convergence with Russia in response to technical requirements that are indifferent to political principles. The process of convergence was carried far beyond the regulation of boiler design when, in the 1920s and 1930s, in the face of violent opposition and controversy, the United States government went into the electric power business. In a surprising dialectic, the divergent tendencies of politically antithetical societies were reversed by the requirements of technology and economics: in the Tennessee valley, electric power was developed on a large scale under public ownership; in the Soviet Union, American engineers, American businessmen, and American private capital, operating through concessions, played a leading role in the development of the electric power industry mandated by Lenin's dictum that "communism is Soviet power plus electrification." An American engineer, Hugh Lincoln Cooper (a Republican), supervised the design and construction of the famous Dnieprostroy in the Ukraine from 1927 to 1932, and the General Electric Company supplied most of the generators for the plant (Dorn, 1979).

By the 1980s hydraulic and industrial intensification had led to a concentration of authority at the center of American society that would have deflated the ego of the most despotic Oriental monarch. In addition to the vast government-owned hydraulic infrastructure created by the Bureau of Reclamation, by the U.S. Army Corps of Engineers, and by the "authorities" that have developed the country's ports and riverine resources, the central government also owns railroad systems, highways, shipyards, and extensive tracts of land; and what it does not own it subsidizes, protects, or regulates.[7] Even in the traditional realms of private capital there is an increasing trend towards government investment. In a recent report to the president, an advisory committee recommended that the United States government establish a multibillion-dollar venture capital fund (*New York Times*, Oct. 30, 1989: D1). Regarding ownership of resources globally, more than 95 percent of the world's oil reserves are held by state-owned companies.

Perhaps nowhere is the federalization of ownership, control, authority, and domination more complete than in the organization of American science. Through its Atomic Energy Commission and its National Institutes of Health, the government owns some of the largest research laboratories in the world. Through the National Security Agency it undertook in 1987 to design and build the largest and fastest supercomputers (Sanger, 1987: A1). It owns high-energy physics research laboratories, the most costly experimental facilities ever built. It is planning an eight-billion dollar superconducting supercollider that the chairman of the congressional Science, Space, and Technology Committee described as "the biggest public works project in the history of the United States" (*New York Times*, Aug. 10, 1987: A13).[8] The space program is financed largely with public funds, the costliest apparatus being government-owned. The National Institutes of Health alone has an annual budget of over seven billion dollars and has more than fourteen thousand employees. Indeed, the bureaucratization of American science has reached the point where government officials could seriously propose that research scientists at NIH wear uniforms in the laboratory (Altman, 1989).

Until World War II and before the revelation that modern science could serve mankind through the development of antibiotics and atomic bombs, the expenditure of public funds on scientific research was small, both in Europe and in North America. Boards of longitude, royal astronomers, national academies and societies, hydrographic and geodetic surveys, botanical gardens, and all the rest had never consumed more than a minuscule percentage of the wealth of any nation. As late as 1930 the total annual cost of research and development in the United States, both governmentally and privately funded, amounted to 160 million dollars, barely 0.2 percent of the gross national product (Ben-David, 1971: 193). Since then the cost of science has reached 150 billion dollars annually (nearly a thousandfold increase), about half of it provided by the federal government alone. In all industrialized nations the social value of scientific research is now recognized, and much of its high cost is covered by public revenues disbursed by the central government. In the ancient East, irrigation networks provided the state with a basis for political domination; in the modern West networks of national laboratories, observatories, university research departments, and federal agencies and commissions serve the same purpose.

The social structure of American science is even more reminiscent of science in the centralized societies of the past than the gross figures suggest. Government support is sharply skewed towards useful knowledge, and pure research is accordingly subsidized at only around 12 percent of the total (Dickson, 1988: 7). And privately supported research, having no obligation of public service, is shifted even further towards the applied end of the research spectrum.

In the Western democracies, especially in their intellectual circles, it is a point of pride that the cultural traditions of the West have sprung from Hellenic roots. With regard to culture in general and to science in particular all roads are reputed to lead to Greece. But the social organization of American science, with 90 percent of it directed towards development and applications, is much more akin to science as it was patronized by emperors, pharaohs, caliphs, and kings than to the purified form that was pursued in Hellenic Greece by individual scholars with no public support. Modern Western governments, like their ancient Oriental counterparts, sponsor the sciences mainly insofar as they perceive them to be useful, while the political authorities of the Greek poleis, in "opposite tension," as Heraclitus might have phrased it, totally neglected the natural philosophy that appeared to them to be so conspicuously useless. Indeed, since World War II the governments of the "developed" world, including the industrial democracies, have not merely employed science; in the name of national security they have demanded the scientist's loyalty to the state above all other loyalties.[9]

While their governments are thus overtly Oriental in their patronage of science, research scientists rejoice in their Hellenic heritage. The physicist Edward Teller declared that pure research "is a game, is play, led by curiosity, by taste, style, judgment, intangibles" (quoted in Reagan, 1967: 1383). And the paleontologist George Gaylord Simpson stated privately, "I am curious in a super-apish way. I like finding out things. That . . . is all that the 'noble self-sacrificing devotion to truth' of 99 44/100% of all

scientists amounts to—simple curiosity" (quoted by Gould, 1988: 15).
Even when scientists are engaged in the most urgent applied research
directed at the most pressing of practical problems, they are easily distracted, if not actually motivated, by "simple curiosity." According to his
biographer, when the late Nobel laureate Albert Szent-Gyorgyi was conducting cancer research, he was in fact engaged in something akin to play
or prayer: "I don't think he was that interested in finding a cure for cancer.
He just got a real charge out of doing basic research. He was devoted to
research the way a hedonist is devoted to pleasure or a religious mystic to
God" (quoted in Johnson, 1988).

Where science is perceived to contribute nothing to material progress
or social well-being, it can make no greater claim to public support than
any other cultural activity that appeals to only a narrow cohort of devotees. Social ambiguity towards science is thus intensified when public
funds are used to sponsor pure research; and heroic, self-satisfied pronun-
ciamentos like those of Teller and Simpson only add to the ambiguity by
strengthening the case of those who would hold scientific research strictly to account in terms of social betterment, national security, or economic gain. Constructing a rationale for the expenditure of government revenues on the portion of research that is not perceived to be useful has taxed
the ingenuity of many apostles of science who have wished to justify the
patronage of pure research. They have generally fallen back on the claim,
securely beyond the reach of empirical falsification, that, however abstract, science will always justify its cost through ultimate usefulness.
Melvin Glashow, a Nobel laureate, has defended the public support of
abstract research on the grounds that "man is curious." But, on second
thought, he and two of his colleagues supplemented that mystical credo
with a utilitarian hope addressed to a practical-minded public:

Why should American taxpayers pay billions so that we who struggle to understand the fundamental nature of matter can learn its most intimate secrets? A
good question, for the high-energy frontier won't lead directly to a cure for AIDS
or an end to the arms race. But our children are ever curious about how the world
works, and it is our duty to tell them as best we can. The triumphs of the supercollider and other scientific initiatives, such as high-temperature superconductors and the space telescope, will inspire more youngsters to become the scientists
and engineers we need to arrest our technological decline and address the tough
questions that the future holds. (Georgi, Glashow, and Lane, 1987)

A more candid and less circuitous justification was offered by Leland
Haworth in 1964 when he was director of the National Science Foundation:

We (the scientists) . . . know the great cultural and intellectual value of science.
But we are not good salesmen. The cultural argument, of course, competes with
similar arguments for other fields of learning. And we would, in my opinion, be

hard put to prove uniqueness for science in this sense. Large federal sums for culture's sake can come only when all culture is heavily supported. So for the present our best drawing card for financial support by the public is the ultimate usefulness of science. I do not defend the fact that this is so; I simply state it as a fact. (Quoted in Greenberg, 1967: 10)

Demythologizing public support for abstract research and presenting less obscure justification for it rests ultimately on clarifying the dual nature of science—its cultural and its practical values, which are generally interconnected and almost always difficult to disentangle. Clarification calls for openly confronting the disjoint loyalties that science generates among individualist votaries enjoying its delights and a public only partially persuaded that practical benefits will somehow result from the most costly and recondite inquiries.

The dilemmas inherent in justifying the expenditure of public funds on the support of pure science were perhaps most vividly revealed in the process of founding and maintaining the government's premier institution devoted to basic research, the National Science Foundation. Immediately after World War II a debate was opened in American governmental circles over the desirability of creating a federal agency to sponsor scientific research. It was recognized that applied science can more readily attract funding from a variety of private sources than pure science, which traditionally had been supported mainly by the universities. Hence, if pure science was to be patronized on a larger scale, it appeared that only the federal government could be its patron. Although there was by then an established belief that pure physics had been instrumental in the conduct and winning of the war, there was persistent opposition to the idea that the central government should sponsor an institution devoted exclusively to pure research. "Basic research versus applied" was, according to J. Merton England, the historian of the foundation, among the "polarizing issues" around which the debate turned. In the end, the question was seemingly settled in favor of federal support of pure science. In 1950, after five years of controversy and debate, an agency was established "dedicated to basic research—research 'performed without thought of practical ends'" (England, 1982: 5–7).

But, appearance was not reality, and the issue was not in fact resolved. The same cultural ambiguity that has always bedeviled the social status of pure research confused the issues (and the congressmen) in the debates over the founding of the National Science Foundation. The enabling act that defined the foundation's objective as the patronage of pure research unfettered by any consideration of practical benefits also specified, without any apparent discomfiture, that the foundation consider the effect of such research "upon industrial development and upon the public welfare" (Lomask, 1975: 30).

The contradictions inherent in the institutionalization of pure science had been foreshadowed in the attitude of Vannevar Bush, the "founding father" of the National Science Foundation. In Bush's view, "support of basic research in the public and private colleges, universities, and research institutes must leave the internal control of policy, personnel, and the method and scope of the research to the institutions themselves" (Bush, 1945: 27). Such an arrangement—institutions connected to government through an infusion of funds but left under their own "internal control"—has been described as "a design for . . . bestowing upon science a unique and privileged place in the public process—in sum, for science governed by scientists, and paid for by the public" (Greenberg, 1967: 106–7). Bush's formula was simply an attempt to square the political circle, the universal desire to depoliticize politics in favor of one's personal interests. "Control of the foundation," including the deployment of its funds, was precisely the "most important" issue that agitated Congress during the debates over the creation of the NSF (England, 1982: 5). The National Science Foundation is the Alexandrian Museum of American science. In both cases, in Ptolemaic Egypt and in the United States in the second half of the twentieth century, exceptionally wealthy states were drawn into the socially ambiguous practice of patronizing abstract knowledge.

At every step the dilemma presented itself, sometimes veiled in confusion. One supporter of the foundation favored an emphasis on pure science but insisted upon "including medicine as a branch, even though medicine was a technology rather than a science." Conversely, biologists preferred "a separation of their fields from the applied science of medicine" (England, 1982: 33). Some congressmen, in the spirit of public servants everywhere, openly found it reprehensible that public funds would be devoted to such a sterile enterprise as basic science and inconceivable that scientific research funded by the federal government would not contribute to the "national defense." In this spirit the president of the University of Notre Dame testified that with World War II out of the way the NSF would be needed in the struggle against communism (England, 1982: 35). There can be little doubt that in general when governments have funded "basic" research they have done so with an eye on applications to statecraft, not only in the exotic worlds of the ancient Orient and medieval Islam, but also today. The pattern may indeed be even more distinct in the modern world, where applied science has demonstrated its superiority over the occult sciences. What government would not prefer to influence the outcome of a war rather than merely predict it?[10] In the course of the congressional deliberations on the desirability of allocating public funds for pure science, considerations of politics and utility were never far from the speaker's podium.

The funding of engineering and social science research raised similar issues. Some of those who took a dim view of science without utility favored the inclusion of the social sciences in the foundation's mission in the hope that, however feeble the social sciences may be as science, they might somehow (at least as their designation suggests) produce social benefits, or, if nothing more, some social awareness (Lomask, 1975: 204–6).[11] The "engineering sciences," despite their unmistakably utilitarian thrust, were included from the beginning among the fields that the NSF supported in its role as patron of "pure" research.

Ironically, the mystification that shrouds the relationship between the state and pure science finally facilitated and expedited the foundation's establishment, in 1950. The confusion and ignorance that surround the scientific enterprise in the public mind and that becloud the distinction between pure and applied research and the relationship between science and society served indirectly to reassure some congressmen that the National Science Foundation would somehow provide a public service:

Probably only a few representatives ever understood it. *If [more] had, the foundation's birth might have been even longer delayed.* The belief that the agency did have a supremely practical purpose—the supplying of scientific and technical personnel and weaponry for national defense—probably did more than anything else to muster the votes needed for passage of the act of 1950. (England, 1982: 7; italics added)

Within twenty years of the foundation's establishment, the pretense that it was designed explicitly and specifically to promote pure science was wearing thin, and it was soon dropped altogether. In 1968 an applied science program was officially implanted in the National Science Foundation under the designation Interdisciplinary Research Relevant to Problems of Our Society (IRRPOS). There are indications that some advocates of the patronage of pure science supported IRRPOS in the hope of fending off more extreme interventions.[12] In any event, amendments to the NSF Congressional Act of 1968 mandated that the foundation's commitment to applied research be expanded. Accordingly, two years later IRRPOS was replaced by "a more elaborate arrangement," and a more permanent one, this time designated Research Applied to National Needs (RANN). The allocation of public funds for scientific research had become a political game, and RANN was a chip that had to be played adroitly. William D. McElroy, who was director of the NSF at the time RANN was created, stated privately that he had an informal understanding with the government that support for basic research would be built up in step with increases for RANN.[13] Indeed, some of that build-up was evidently accomplished by subverting RANN's mission and reallocating some of its funds to pure research. Five years after the creation of RANN it was estimated that "about 40 percent of the funds allotted to the program are spent on the search for new scientific knowledge" (Lomask, 1975: 244). RANN could never be genuinely acceptable to a scientific community whose statesmen see science as the descendent of Greek natural philosophy and whose most eminent members are renowned for extending the frontiers of abstract theoretical knowledge. A typical RANN project, such as research on an aerodynamic windbreaker to reduce the gasoline consumption of long-distance trucks, could never grip the imaginations or command the loyalties of the scientists and scientist-bureaucrats who were drawn to the NSF by its reputation as a patron of pure science.

Bureaucratic combat over public patronage of science is conducted on many levels and employs a variety of tactics, some of them devious. As a result of the confusion that obscures the distinctions between pure and applied research, along with the ambivalence in the objectives of government patrons, applicants for funds are subjected to unsavory temptations and provided with convenient subterfuges. When RANN appeared menacingly on the horizon, NSF functionaries set up a new unit, the Division of Institutional Development, which succeeded in "so revamping a number of requests for institutional grants that they could be accepted and funded as problem-oriented projects" (Lomask, 1975: 244–45). It is common, albeit rarely publicized, knowledge among scientists that such benign deceptions are widely practiced. Occasionally they are frankly admitted, since they are hardly considered improper. One scientist-bureaucrat stated:

I came to NSF out of NASA where we did one hell of a lot of fundamental research under the cloak of applied science and applied engineering. We supported the whole Goddard Space Flight Center (home of a variety of basic-science activities, ranging from solar physics to intensive study of fields and particles) . . . under the mantle of the applications program. (Quoted in Lomask, 1975: 242)

If pure research were provided with public revenues on a modest level, comparable to the support allocated for other purely cultural activities, justifications like "man's curiosity" might suffice; but high-energy physics alone, which Melvin Glashow freely concedes neither offers nor promises any direct benefits in technology or medicine, requires billions of dollars for particle colliders and other apparatus. On that level of expenditure, the state is expected to apply a more socially rational accounting principle, and it remains far from clear what that principle may be—unless it is technological spinoff, however remote. Despite even the disclaimers of high-energy physicists themselves, bureaucrats, public servants, and elected representatives evidently continue to believe in the utilitarian potential of a field with as suggestive a designation as "high-energy physics," with procedures that to the layman seem so reminiscent of the Manhattan Project of World War II, and which the government of the Soviet Union so eagerly supports. What other motive would induce the American public and its representatives to spend billions on a single field of pure research? Moreover, if man's curiosity were in fact the guide in the costly quest for knowledge, how is a three-to-one partiality towards physics over chemistry to be rationalized? (Greenberg, 1967: 11–12). And why is paleoanthropology offered only a pittance? Surely man is as curious about his own origins as he is about the structure of matter.[14]

These questions are relevant only where public funds are used to support pure science, and such allocations are made only where governments are bureaucratically centralized. In the city-states of Greece, among hunter-gatherers, in Neolithic villages, in societies of pastoral nomads, and in feudal fiefdoms, no central authority ever patronized science. But in the industrial civilizations of the twentieth century, as in the ancient Oriental world and the medieval East, higher learning, arcane and filled with utilitarian promise, has been enlisted in the service of the state and society. In such service the individual scientist's interest in abstract inquiry has always been confronted with cultural ambiguity and searching scrutiny.

Despite our high regard for the traditions of the Hellenic enlightenment and the intellectual triumphs of pure science, it is the traditions of the East as much as those of the West that we copy in the actual practice and organization of scientific research. The scribal schools of ancient Egypt and Mesopotamia, the Astronomical Bureau of China, the Alexandrian Museum, the Byzantine hospital, and the Islamic *madrasa*, no less than the natural philosophy of Ionia and classical Athens, are our models. In the value we place on pure research, we are Greek. In our institutional patronage of science, we are Babylonian.

TOWARDS AN ECOLOGICAL MODE OF PRODUCTION

Marxists and non-Marxists alike, for incongruent but overlapping reasons, reject "technological interpretations" of history. Philosophically, a technological interpretation is often rejected on the ground that it reduces an account of history to one or only a few causal factors and diminishes the scope of human action and volition. These objections are easily overdrawn. Emphasizing the importance of agricultural and industrial processes and analyzing the role they play in shaping historical events can enrich the study of history as readily as they can impoverish it, depending only on the craft of the historian. That approach is no less humanistic than more orthodox social or political history. Contracting with the Earth an exchange of work and worry for food and fiber is as humane an act as devising a constitution; and studying the technical terms and conditions of that contract is no less humane. It can properly be placed beside the social contract without diminishing or defiling the study of history. The histories of technology and agriculture, when closely interlaced with other historical processes, may expand, not compress, the explanatory repertoire of historical analysis by bringing into focus developments that are often neglected. They can, moreover, direct attention to material forces that have often been seen as irrelevant to social development but that may at times totally override the effects of social and economic relations and even of whole modes of production. Productive forces include more than raw materials, machines, technique, and knowledge. They must also include ecological factors—water, air, soil, conditions of weather and climate, and demography. Writing in the mid-nineteenth century, Marx steadily overlooked these forces of production, which must have seemed at the time to be inexhaustible and to be changing so glacially that they were effectively outside of society and history.

One hundred fifty years later, the world is ominously different. Accessible fresh water, which even to begin with amounted to only one-hundredth of one percent of the planet's total water resource, is now, under demographic pressure and qualitative degradation, in short supply. Topsoil continues to be flushed into streams and carried out to sea. The ozone layer, of which Marx was blissfully unaware, has been pierced over the Antarctic and is decreasing globally (Bowman, 1988), confronting all societies, whatever their putative modes of production, with the most lethal prospects. And a global warming trend has become sufficiently threatening to lead an official of the United States government to confess that it is one problem whose solution cannot be found in the corridors of Washington or even of the free world, and "must be addressed internationally" (Shabecoff, 1986).

It has become much more disagreeably clear since the nineteenth century that geographical variables operate on historical time scales, and may at times overwhelm all other variables. The unshakeable durability of the Asiatic Mode of Production over long periods of time might already have convinced social analysts that ecological constraints can be overpowering. Six thousand years ago ecology molded social development into a characteristic mode of production—hydraulic civilization. Now, at the end of the twentieth century, accelerating destruction of the environment is swamping some of the issues and conflicts that have tended to produce divergent modes of production and is generating convergent social forces as industrialized nation-states slowly and unwillingly close ranks in the face of common dangers. If the ancient East can serve as a guide, industrial civilization can define a relatively stable mode of production only if it comes to terms with its physical environment without regard to divisive ideologies.

But lessons are reluctantly drawn from a way of life so remote and strange as the Asiatic Mode of Production. Historians have frequently shown a disinclination to derive meaning from "stagnant" social systems like food-collecting bands, pastoral nomadism, and the Oriental societies of the ancient East. Too often these systems, having produced relatively few documentary records, are regarded as more appropriate for anthropological or archaeological, rather than historical, study. But it is in precisely these societies, if only in their repeated formation across the world and across time and in their stolid immutability, that ecological factors can best be seen to have played a most commanding role.

Environmental concerns are slowly increasing the awareness among scholars that ecological processes, more frequently than has commonly been believed, can be as historically significant as political processes. The recognition that physical forces can drown out the differences between social systems has even been readmitted into the communist world, from which it was officially expelled sixty years ago when the Asiatic Mode of Production was proscribed as a historical concept. The president of the USSR, Mikhail Gorbachev, has called for a new emphasis on "ecological security," and he recently stated that: "for all the contradictions of the present-day world, for all the diversity of social and political systems in it, and for all the different choices made by the nations in different times, this world is nevertheless one whole. We are all passengers aboard one ship, the Earth, and we must not allow it to be wrecked. There will be no second Noah's Ark" (Gorbachev, 1987: 12). And in another display of convergent values, Gorbachev's environmentalism was matched by the president of the United States, George Bush: "No line drawn on a map can stop the advance of pollution. Threats to our environment have become international problems. We must develop an international approach to urgent environmental issues" (New York Times, Sept. 26, 1989: A16).

In similar environments allegedly divergent modes of production become in fact soft, shifting, and overlapping categories. Even in more systematic sciences, taxa often shade into one another in defiance of taxonomy. Taking Marx's material mode of production as the criterion, the United States and the Soviet Union are in fact substantially indistinguishable. If hand mills and steam mills "give you" lords and capitalists, in Marx's formula, what do you get from hydroelectric and atomic-electric stations, gas turbines, uranium ore, semiconductors, and industrial robots, all of them as characteristic of the Soviet as of the American economy? Viewed in geographical perspective, socialism and capitalism are not alternative modes of production; the processes of production are essentially the same in both social systems. Rather, they are alternative paths through the transition from agricultural to industrial civilization. During the transition, one social group or another may be favored by the chosen path; but once the transition is complete, the paths converge under the convergent pressures of industrial intensification in similar ecological and technological environments. "Socialist" economies must embody market mechanisms just as "capitalist" economies have embraced state ownership and control.[15]

And what kind of society does contemporary science, similar everywhere, "give you"? Of all cultural spheres, contemporary scientific knowledge is the most rigidly universal, indifferent to all distinctions, including modes of production. The arts, the humanities, and the social sciences march to different drummers in different cultural settings, but calculus and thermodynamics are smartly in step everywhere. Not only the intellectual and technical content of science, but also its social organization are at present everywhere similar, at least insofar as governmental patronage of science is as heavily skewed towards applied research as it is in both the Soviet Union and the United States.

At the end of the twentieth century the polarizing claims of ideologically defined modes of production have been refuted by the reality of an industrial mode of production on a global scale. Insofar as the relationship between science and society can be clarified by reference to geographical conditions and ecological habitats, the polarity that can be most enlightening is, instead, that between the Asiatic Mode of Production and Hellenic Greece—the principle of the bureaucratic institutionalization of science in opposition to the unfettered inquiry of individual scholars pursuing their abstract studies wherever they can find the sociological space to indulge their idiosyncratic curiosities. That unstable opposition between the individual's passion to know and the community's need to harness higher learning to the wheel of social betterment has been a powerful motor of scientific progress.

NOTES

INTRODUCTION

1. Cf. Hartshorne (1939), x: "The use of the geographic approach in historical studies, in particular the attempt to determine the significance to history of specific geographic features is logically a part of history—whether pursued by historian[s] or geographers."

2. Cf. Gaile and Willmott (1989), xxvii: "Despite the aggressive search for new directions, geography remains firmly grounded in its traditional concerns. Location and distance remain important. Sense of place remains important. Interaction between nature and society remains important. Regions, too, are still major organizing frameworks. Maps . . . are still a basic means by which geographic information can be conveyed."

3. Although the concept of *lebensraum* was given a menacing twist by Nazi ideology, in Ratzel's usage it is probably best translated simply as "habitat" (see Hunter, 1983: 239–40).

4. Semple was regarded as one of Ratzel's most brilliant students, the only woman among several hundred. At the time that she studied with Ratzel at Leipzig in the 1890s she was not permitted to matriculate, and to attend his lectures she was required to sit alone in an adjoining room. She went on to become president of the Association of American Geographers.

CHAPTER ONE. SCIENCE AND MATERIAL LIFE

1. Karl Wittfogel (1981b: 374) credited the geographical explanation to Marx without mentioning Engels' priority.

2. This materialist scenario may be highlighted by contrasting it with M. I. Rostovtzeff's mentalist explanation of the durability of the Bronze Age kingdoms: "The Oriental civilizations were more stable and lasting than the Greco-Roman, because, being chiefly based on religion, they were nearer to the masses" (Rostovtzeff, 1957, vol. 1: 541).

3. The quoted term was used by Marx in 1853 (Marx & Engels, 1980: 35).

4. He added sardonically: "It was reserved for the enlightened English to lose sight of this in India; they let the irrigation canals and sluices fall into decay, and they are now at last discovering, through the regularly recurrent famines, that they have neglected the one activity which might have made their rule at least as legitimate as that of their predecessors."

5. Marx cited the French zoologist Georges Cuvier (1769–1832) as the authority for the astronomical part of his opinion.

6. It may be relevant, in accounting for the interests of American anthropologists and archaeologists, that in South, Central, and North America, all of the ancient high civilizations that they study depended on irrigation or some other intensive form of hydraulic management of agriculture.

7. For a concise and temperate review of the controversy, see Mitchell, 1973.

8. Over the past thirty years the bibliography on the subject has become a divergent series. Accounts of the renewed discussion may be found in many publications including: Lichtheim, 1963; Pečírka, 1964 and 1967; Skalník & Pokora, 1966; Danilova, 1971; Krader, 1975; Hindess & Hirst, 1975; Bailey & Llobera, 1981; Dunn, 1982; Fogel, 1988; and O'Leary, 1989.

9. The unsigned essay "Dialectical and Historical Materialism," from which these passages are taken, has been credited to Stalin. Some have questioned the attribution. It should be noted that Stalin was treating geographical factors as temporal processes and therefore stressed the importance of the pace of change; however, in sociological and historical discussions it is generally synchronic differences between regions that are considered.

10. It may be noted that those who condemn many "Asiatisms" often recognize many "feudalisms." On the promiscuous use of "feudalism" by historians, cf. Simon (1962), 184: "Obviously a socio-economic stage which covers both Ruanda-Urindi today and France in 1788, both China in 1900 and Norman England, is in danger of losing any kind of specific character likely to assist analysis."

11. Cf. Lawton and Wilke (1979), 3: "One observation that emerges from a review of the works cited . . . is that a surprising number of early agricultural economies, both those of independent origin (such as in Peru, the Southern Mexican Highlands, and the Near East), and those that were recipients of farming technology diffusing from such centers of origin (including the American Southwest and the Nile Valley), developed in drier regions of the world."

12. Wittfogel's definitive study, *Oriental Despotism*, which appeared in 1957, led one reviewer to grandiloquent praise for the "light which [it] throws upon the nature of Soviet totalitarianism, which stands revealed as one of the most tragic retrogressions in the history of mankind." He added, "This [book] may well be our most effective intellectual answer to Communism as it operates in today's world" (Walker, 1956–1957: 603).

13. For a systematic discussion of these views, see Mark Nathan Cohen, 1977. Cohen's conclusions are briefly summarized in Diamond, 1987: 64–66.

14. The identification of science and the crafts in the prehistoric and ancient eras is a common mistake made not only by Marxist writers. See, e.g., Lloyd, 1970: 2–4.

15. For another case of the coincidence of unlike frontiers, see Owen Lattimore's essay "The Frontier in History": "The frontiers between types of soil, between farming and herding, and between Chinese and Mongols coincided exactly" (Lattimore, 1962a: 473). We are reminded by William McNeill that it was "narrow-minded political persecution [that] seems to have buried Owen Lattimore's elegant essay" (McNeill, 1983: 63).

16. It was long believed that Stonehenge could only have been the work of Bronze Age contractors who migrated westward from Egypt or Mesopotamia, but

that opinion has been discredited (Daniel, 1980: 78–85). Current research indicates that metal working and the presence of full-time craft specialists, evidently capable of constructing megalithic monuments, were features of an indigenous "high" Neolithic in the Near East as early as 7,000 B.C. (Stevens, 1986: C1). If, as Colin Renfrew has proposed, the British Isles and Western Europe were peopled by Indo-Europeans from the Near East, Stonehenge would have been the work of those people, and they would have been in possession of an adequate technology: "The Celtic language would have evolved in western Europe from Indo-European roots. Rather than an alien group obliterated by the Indo-Europeans, the people who built Stonehenge and the other great megalithic monuments of Europe were Indo-Europeans who spoke a form of Indo-European ancestral to the Celtic languages of today" (Renfrew, 1989: 113).

17. It should not be assumed, however, that it is only the mobility of nomads that prevents them from cultivating science. They are evidently capable of practicing prescientific metallurgy on the move. See Lattimore (1951), 329: "Nomadic life by no means prevents the working of metal on a small scale by individual craftsmen. The necessary forges, tools, and supplies of raw material can be carried about."

18. Dryland (or dry) farming is dependent entirely on moisture from rainfall or runoff from snowfall at high altitudes, without any attempt to impound runoff or employ irrigation techniques.

19. The extent and social significance of artificial irrigation in ancient Egypt have been called into question in a disputed (albeit authoritative) opinion by Karl Butzer (1976, esp. 109–11). For a concise review of the issue along with a counterargument, see M. Harris (1977, 241–45). Worster (1985, 43–44) reached the same conclusion as Harris.

20. The rivers of Mesopotamia carried silt loads four or five times as great as the Nile. They drain territories that Neolithic farmers had heavily deforested and that could not, therefore, hold their soil. In contrast, the Nile, at least the White Nile, rose in regions populated only by hunters and gatherers who had left the pristine forests intact. (See Lawton & Wilke, 1979: 10–11.)

21. For the argument that some of the architectural building projects, particularly the pyramids, were calculated to keep teams of corvée labor employed in the agricultural off-season, see Mendelssohn, 1970: 215.

22. Among physicians, there were also those, in Egypt at least, who were employed privately. Paul Ghalioungui has concluded that "many physicians were salaried by the state or by private persons." He also showed that there was a category of physician-scribe in Egypt (Ghalioungui, 1983: 97, 101).

23. "The high regard of officials for their scribal activities and training was also reflected in the fact that a great many of the prominent officials during the whole of Egyptian history had statues of themselves sculpted in the scribal position, with knees crossed and papyri on their laps" (Clagett, 1989: 2). In referring to "nobles and high officials" Clagett states (p. 12) that "most of [them] started as scribes."

24. Cf. S. Kramer (1959), 1: "The vast majority [of cuneiform tablets] are administrative in character; they cover every phase of Sumerian economic life." And

Clagett (1989), 2: "To understand why the scribe was an important person in ancient Egyptian society, we have to realize how much the highly organized state and its culture depended on the invention and development of writing."

25. For this reason the *DSB* abandoned the biographical approach in dealing with the Oriental and ancient American civilizations and covered them with topical articles instead (see Gillispie, 1970–1980: vol. 15). Similarly, Gascoigne (1984), which is "a chronological list of 13,300 persons who were of some degree of significance in the development of science," also contains no individuals prior to Thales.

26. For the architects, see Sarton (1952), 32: "The Egyptian architects and their royal masters were fully conscious of their achievements and proudly recorded them. Half a dozen obelisk architects are personally known to us."; see also p. 34. Ghalioungui (1983: 16–37) identifies around 150 pharaonic physicians.

27. This conjecture is strengthened by the fact that for Egypt the secrecy of scribal learning is well-attested. (See Clagett, 1989: 26, 28, 31, 32, 187, 206.)

28. For a discussion of some of these developments in the context of the Babylonian tradition of celestial divination and calendrical calculations, see Toulmin and Goodfield, 1965: 23–40.

29. The Egyptians used a recipe that was equivalent to the correct equation:

$$V = h/3 \times (a^2 + ab + b^2).$$

The formula can be rigorously derived only by using the theory of limits. It is unknown how the Egyptians arrived at their procedure.

30. The interested reader may wish to follow the mathematical procedure of the ancient scribes:

> If it is said to thee, a truncated pyramid of 6 cubits in height,
> Of 4 cubits of the base by 2 cubits of the top,
> Reckon thou with this 4, its square is 16.
> Multiply this 4 by 2. Result 8.
> Reckon thou with this 2, its square is 4.
> Add together this 16, with this 8, and with this 4. Result 28.
> Calculate thou $\bar{3}$ [i.e. 1/3] of 6. Result 2.
> Calculate thou with 28 twice. Result 56.
> Lo! It is 56! You have correctly found it.
>
> (Gillings, 1978: 697)

31. Often, however, the lists were compiled with practical, albeit magical, intent—to bring into being the things listed. Cf. Clagett (1989), 239: "I should stress that often the lists of names are given with the express purpose of bringing into existence that which is named."

32. Astronomy and its corollary, political astrology, may have been more generously patronized in Babylonia than in Egypt because of its strategically precarious situation.

33. The designations are attributed to J. H. Hexter (Kennedy, 1985: 33).

34. The racial purity of the Greeks is curiously beside the point. Even if they had been a hybrid population, Heath's assertion would be baseless. It must be conceded, however, that while ethnic explanations of social and historical events are unwarranted at present, they cannot categorically be ruled out of consideration.

Cf. Cipolla (1970), 12: "We are in complete ignorance of the interaction between cultural and biological development and, unfortunately, not enough scholars dare to venture into this essential field for fear of being accused of racialism, nazism, or some other kind of unpleasant mental disposition."

35. Ben-David was surely mistaken in his claim that there were "natural philosophers" in the ancient Near Eastern civilizations or that any groups in those civilizations played a role similar to Hellenic Greek scientists. See Kuhn (1972), especially p. 169, for criticism of Ben-David's study. Also, cf. Clagett (1989), 263: "Even the briefest glance at the extraordinary volume of ancient Egyptian literature that touches upon the nature of the world and how it came into being will convince a reader that during the three thousand years of Pharaonic Egypt there was no natural philosophy or physics that was separate from religion, myth, and magic."

36. That geography enters into the social and political pictures is sometimes conceded. Cf. Lloyd (1979), 241: "The formation of small independent political entities was, no doubt, favoured by geographical factors in the Aegean area." For a tentative but aborted approach to a geographical interpretation of the history of ancient science, see Olson, 1982: 17–19.

37. Plato was evidently sufficiently wealthy to establish and endow his famous Academy in Athens so that it could operate without the benefit of tuition fees or government support (Cameron, 1969: 11).

38. It may be that arithmetic is initially more useful in practical affairs than geometry. In Babylonian mathematics algebra was well developed as an independent science while geometry was merely a field of application of numerical and algebraic methods. Cf. Neugebauer (1959), 45, on Babylonian mathematics: "The mathematical importance of a problem lies in its arithmetical solution: 'geometry' is only one among many subjects of practical life to which the arithmetical procedures may be applied. . . . ['Geometry'] is no special mathematical discipline but is treated on an equal level with any other form of numerical relation between practical objects."

CHAPTER TWO. SCIENCE IN THE ASIATIC MODE OF PRODUCTION

1. The free-thinking Soviet historian-anthropologist A. M. Khazanov (now an émigré) denies that pastoral nomadism is a mode of production; he places it "in one typological line alongside agriculture, hunting and fishing." While he rejects "exclusively" ecological explanations of sociopolitical organization, he agrees that "of course, the ecological factor is one of the most important determinants" (Khazanov, 1984: 192–93).

2. The European agricultural revolution is discussed below in Chapter 5 under the rubric "New Trends in Cultivation."

3. The historical pattern of agricultural development in Ceylon seems to have been a diffusion of agriculture from the wet zone to the dry zone, a pattern that has recurred two or three times in the history of Ceylon and is again under way today as the dry zone is being redeveloped and resettled (Fernando, 1982: 43–44).

4. Cf. Fernando (1982), 27: "What is most amazing is that major and medium ancient hydraulic works are exclusively in the Dry Zone."

5. The correlation between monumental architecture and the rise of advanced agricultural societies has frequently been noted. Indeed, such architecture has been described as "a diagnostic of nonegalitarian societies" (Barbara Price, 1978: 165). It should be noted, incidentally, that heavy construction in Ceylon was facilitated by the strength and intelligence of the elephant.

6. On the similarities between Ceylon and Cambodia, cf. Leach (1959), 6: "Burma (Pagan), Cambodia, and Dry Zone Ceylon (Sinhala) were all sophisticated states of relatively small scale for which we possess a fairly detailed political history; they all flourished at much the same period; they were all 'hydraulic societies' and they were all in regular contact with China."

7. Even today, "India's traditional faith in astrology" performs a state function. An article in the London *Financial Times* reported that: "this astrological forecast has been made by an official responsible for water supply, Mr. Vinod Pande—a distinguished civil servant from the finance and commerce ministries, and secretary for rural development. Mr. Pande's colleagues were dismayed when he told them last month that it would not rain significantly in the dry areas until . . . Jupiter became retrograde, having been directional, and Saturn did the reverse. But they were glad, and India's traditional faith in astrology was boosted when, on cue, it poured that afternoon in Delhi" (Elliott, 1987).

8. Leach noted a "government 'irrigation department' of some sort" but dismissed it as insignificant (Leach, 1959: 26).

9. Both the degree of centralization and the intensity of the tyranny may vary as much as the scale of the engineering. According to Needham, in China the toil of the "unnumbered millions of men and women" that produced the hydraulic works was "more often willing than otherwise" (Needham, 1971a: 378). With regard to ancient Ceylon, cf. Fernando (1980), 4: "The people participated actively, as they inherited the product of their labour."

10. Wittfogel's insistence, in some versions of his thesis, on the unremitting despotism of hydraulic civilizations may be untenable, but in fact he displayed less consistency than Leach gave him credit for. In an early work (Wittfogel, 1938: 99) he even allowed for a "feudal" tendency in situations where irrigation was too localized to result in strong central control.

11. In ancient China, the modern word for astronomy designated astrology and the discernment of celestial portents.

12. The introduction of listmaking is attributed to Tsou Yen (fourth to third centuries B.C.), "the real founder of all Chinese scientific thought" (Ronan, 1978: 142).

13. On the question of whether the Maya invented the zero independently or derived it from the Old World, see Seidenberg, 1986.

14. Cf. Omohundro (1987: 14): "Handmade beds and fields do consume much labor. Building some of the larger complexes, like those in Mexico, also requires a high degree of coordination among farmers." (See also Turner, 1974, and Chen, 1987.) Intensive raised-field agriculture was also practiced in ancient Peru (Stevens, 1988: C1, C15).

15. *Pueblo* here refers generically to the various pueblo-dwelling groups who have inhabited the southwest region since prehistoric times.

16. In response to petitions by the native inhabitants of the region, Snaketown

has been made a national monument; it was incorporated into the National Park System in 1972.

17. This account of Hohokam prehistory is drawn mainly from Haury's report (Haury, 1976).

18. For the opinion that the collapse of Hohokam civilization resulted from the ruination of its agricultural base, see Worster (1985), 34: "Intensive irrigation has everywhere led to increasing concentrations of salts in the topsoil, poisoning the farmer's fields. That nemesis came to the Hohokam too, and they were forced one day to abandon their agriculture completely, leaving behind them whited fields and dust-drifted canals."

19. For the concept of ecological imperialism, see Crosby, 1986.

20. Cf. Hunt and Hunt (1978), 78: "We have to talk about differences in the degree of centralization rather than its simple presence or absence."

CHAPTER THREE. SCIENCE AND CULTURAL AMBIGUITY

1. Cf. Hyams (1976), 55–56: "The Nile Valley soil has proved resistant to seventy centuries of what can only be called abuse. . . . The soil of the Nile Valley is probably the only soil in the world which, by virtue of annual renewal combined with the unique topography of the country, has turned out to be nearly manproof."

2. Although there were far fewer irrigable fields than in the great river valleys, irrigation in the Greco-Roman world is confirmed by a developed terminology and by venerable irrigation laws and customs. The sting of the term *rival* is suggested by Ellen Churchill Semple's aptly phrased observation: "Such an irrigation channel was called a *rivus*, and all persons participating in its water rights were *rivales* or rivals, a term which came to be applied to rivals in love by the old Roman comedies,—with peculiar appropriateness, for competition in water rights was keen, equity in them was jealously guarded, illicit draughts were common, and quarrels were frequent" (Semple, 1931: 466).

3. Although prior to the advent of agriculture Greece was almost completely covered with trees, today less than 10 percent of it is forested, and less than 15 percent of it is arable. It is ironic that deforestation is initiated in order to increase cropland.

4. On the Greek timber famine, see also Rostovtzeff, 1941, vol. 1: 91, and Semple, 1931: 267, 275ff.

5. While conceding that "the mountainous terrain of Greece no doubt favored the establishment of small political units," McNeill (1963: 194) has argued that the correlation between geography and political compartmentalization was imperfect, and that other divisive factors must have come into play. However, with regard to the "peripheral relation" of Greece "to the center of Middle Eastern civilization," he added: "In this sense, perhaps, geography had more to do with the success of the Greek polis organization than I have allowed in the text."

6. Plato's Academy and Aristotle's Lyceum were only private schools that perpetuated the thoughts of their founders.

7. For a clear description of the Marxist concept of the cultural superstructure, see G. A. Cohen (1978), 45: "The superstructure consists of legal, political, re-

ligious, and other non-economic *institutions*. It probably includes universities, but it does not include knowledge, for knowledge is not an institution."

8. Joseph Needham attended that London conference, and Hessen's paper reinforced his interests in both the history of science and Marxism. It is worth recalling Needham's memorial to Hessen and some of his co-delegates (including Bukharin): "It would hardly be possible . . . to ignore the tragic fact of the disappearance of so many of these delegates in the years after the Congress, according to the dreadful principle that 'all revolutions devour their own children'" (Needham, 1971b: vii–ix).

9. A more extensive account of the procedure is in Merton, 1970: chapter 10.

10. The biographical data used to make the counts in this section were drawn from the *DSB*. It was necessary to rearrange the entries chronologically; and in order to assign dates, the age of forty was selected to mark the prime of these scientists' careers.

11. Plato's Academy and Aristotle's Lyceum were not institutions, since they had no commitment to public service. They were essentially private clubs.

12. Farrington's thesis that the Milesians derived their ideas from the technical basis of their society is called further into question by the possibility that some of those ideas, specifically the theory of elements (earth, air, fire, and water), may have been diffused into the Ionian confederation from Persian communities to the east (Sarton, 1959: 6).

CHAPTER FOUR. FROM PHARAOHS TO CALIPHS TO LORDS

1. It is preferable, following George Sarton, to retain the Greek form of the name in order to distinguish the governor from the famous Alexandrian astronomer, Ptolemy (100 – c. 170).

2. It was contested in size of population only by Patna, on the Ganges (Chandler & Fox, 1974: 80).

3. The wealth of Alexander's eastern conquests was not limited to Egypt: "There is no reason to suppose that Syria proper, Mesopotamia, and Babylonia, with their great agricultural resources, were less thriving than the Ptolemaic dominions" (Rostovtzeff, 1941, vol. 1: 536–38).

4. Cf. Tarn (1975), 269: "Above all, the kings established libraries in their capitals. The idea of a library had probably come down from Assyria and Babylon, but in the Greek world before Alexander only an occasional tyrant had had the money to collect books. . . . State libraries now appeared at Antioch and Pergamum, and later at Rhodes, Smyrna, and probably other cities."

5. In 1966 the U.S. Department of Defense published the results of an elaborate twenty-year study, Project Hindsight, which indicated that only 9 percent of the technological innovations that were studied had sprung directly from science, pure and applied, and only 0.3 percent were the product of pure science alone. The results "came as something of a bombshell to the scientific community" (Layton, 1971: 563–64).

Although Nobel Prize awards are generally assumed to be a measure of a nation's success and strength in scientific research, an American congressional task force conducted a study of the correlation and concluded that Nobel Prize research is either unrelated to useful results or, in the case of physics and chemistry, is *negatively* related to practical benefits: "There is no relationship between national Nobel-Prize winning in physiology or medicine and public health status, and there is actually a negative relationship between national economic performance and the hosting of Nobel-Prize-winning research in physics and chemistry" (Congressional Research Service, 1986: xiv).

6. George Sarton commented on the anecdote: "There are still many idiots today who would judge education as Euclid's student did; they want to make it immediately profitable, and if they are given their way, education vanishes altogether" (Sarton, 1959: 36).

7. Those districts had been rich sources of exports of grain, fish, and timber as far back as the Archaic period of Greek civilization (Boardman, 1963).

8. Some have suggested that the population decline of the west Roman Empire was a result of the debasement of Roman agriculture (Carter & Dale, 1974: 142–43). Rostovtzeff (1957, vol. 1: 376–77) rejected the claim. That agricultural degradation can in principle be an "important" factor in the decline of a whole civilization was concluded by Jacobsen and Adams (1958: 1252), on the basis of archaeological evidence, with regard to Sumer: "Probably there is no historical event of this magnitude for which a single explanation is adequate, but that growing soil salinity played an important part in the breakup of Sumerian civilization seems beyond question."

9. Cf. also Teall (1959, 91): "a steady, predictable supply of grain from Egypt was absolutely essential to satisfy the needs of Constantinople and the expeditionary army in the eastern provinces."

10. Based on a count of entries in the *DSB*.

11. Trofim Lysenko was a Soviet plant breeder who advocated the principle of the inheritance of acquired characteristics and rejected the principles of genetics. He won the endorsement of Communist Party authorities in the 1930s; and, as a result, genetics research was prohibited in the Soviet Union for more than twenty years.

12. "Indeed, Moscow probably has the greatest concentration of mathematical talent of any city." "Today the Soviet Union has the world's largest community of scientists and engineers, exceeding that of the United States by almost a third." (Graham, 1989: 3, 1)

The first cyclotron in Europe was constructed in the Soviet Union in the 1930s. "By 1941 they were erecting two more of these gigantic atom-smashing machines. One was designed to attain a radiation strength three times as great as that of the biggest apparatus then operated in the United States" (Jungk, 1958: 263).

13. Even in the extreme case of Nazi Germany, when some Nazi physicists, claiming the superiority of "*volkisch*" over "Jewish" physics, appealed for the support of the political authorities, they were told by Nazi officials that the issue was internal to physics and must be resolved by the scientists themselves. "Nazi staff members saw the Aryan physics issues as an internal dispute within the physics discipline, and they did not feel competent to judge the technical matters under debate. Inability to convince these officials that their struggle had practical consequences was one of the key failures of the Aryan physicists" (Beyerchen, 1977: 205).

14. Cf. Cameron (1969), 12: "As late as *ca.* 560 the confiscation of the Academy revenues supposedly confiscated in 529 was *still* not complete. . . . Evidently the authorities at Athens were not very energetic in carrying out Justinian's orders."

15. In the sixteenth century, Philip II's Spain was governed under a "state religion," and Protestant teaching was suppressed in the universities—but Copernicanism, despite being scientifically deviant and theologically tainted, was tolerated (Goodman, 1988: 52).

16. According to one authority it was in fact a totally insignificant event: "Scholars have credulously regarded the laws of 529 [closing the Academy] as the final victory of Christianity. Romantics have lamented the final suppression of the 'Greek Spirit.' In fact, most late Roman laws prove little more (in the words of A. H. M. Jones) than that 'the abuses which they were intended to remove were known to the central government.' After a few months nobody took any more notice of this law than of any other" (Cameron, 1969: 9).

17. Based on entries in the *DSB* and in Gascoigne (1984).

18. A fresh study of the history of Sassanid science and medicine is long overdue.

19. Perry Anderson refers repeatedly to "Muslim disdain for cultivation," Islamic "indifference or contempt for agriculture," "the economic spoliation of agriculture in the great Islamic Empires," and the Islamic "contempt for agrarian production" (Anderson, 1974: 499, 501, 505). This historiography of Islamic agriculture has been discredited (see Watson, 1983: esp. 5–6). Although Anderson follows in Marx's footsteps he could not have derived his views of Islamic agriculture from the master: "The secret of the flourishing state of industry in Spain and Sicily under the dominion of the Arabs lay in their irrigation works" (Marx, 1936: 564).

20. "Without improvements in the extent and quality of irrigation, the new agriculture could not . . . have been diffused on a significant scale. . . . A leading role was played by the State" (Watson, 1974: 13, 27).

21. Even as late as the fifteenth century the towns of Europe had small populations. Cf. Pirenne (1937), 170: "Medieval towns were only thinly populated. Strange as it may seem, it has been proved that Nuremberg in 1450 numbered only 20,165 inhabitants; Frankfort in 1440, only 8,719; Basel about 1450, round about 8,000; Friburg in Switzerland, in 1444, only 5,200; Strasburg, about 1475, only 26,198; Louvain and Brussels in the middle of the fifteenth century only about 25,000 and 40,000 respectively."

22. Based on a count of entries in the *DSB*. By comparison, during those four centuries after Muhammad (700–1100) the Latin West produced fewer than one-sixth the number of scientists that practiced in the Islamic lands.

23. Nasr (1968), 188: "The link between Islamic medicine and the older schools is to be found in the school of Jundishapur, which must be regarded as the most vital organic connection between the Islamic and earlier traditions of medicine."

24. Shiite institutions in particular leaned towards secular studies, including fields of Greek science (Nasr, 1968: 71–77). For the argument that the college, as distinct from the university, was an Islamic innovation, see Makdisi (1990).

25. Cf. Sayili, 1981: 225: "It is certain that in Islam royal initiative and patronage were factors of primary importance in the construction of observatories."

26. It may be noted, in further refutation of the thesis that "state religions" suppress science, that in Islam the reverse was true.

27. Astronomy sank deep roots in Islamic culture. V. S. Naipaul has recorded the following impression of an ayotollah he met in Iran in 1979: "He looked like what he was, a figure of high medieval learning. Philosophy and astronomy had been among the subjects he had studied in Qom in the 1920s under a famous divine: astronomy part of the Muslim intellectual expansion of centuries before, but long since frozen, with philosophy, into a theological discipline" (Naipaul, 1981: 70).

28. The discussions were published as Part 2 of Coulborn, 1965.

29. Cf. Sayili (1981), 397: "[Is] the fact that Eastern Islam had clearly a much more important part in the evolution of the Islamic observatory as compared with the Maghrib and Spain . . . in any degree explainable through the hypothesis of Fareastern influences?"

30. On the strength of these ecological differences, which continue to divide the world, Kipling could just as well have written: "Oh, North is North, and South is South." Cf. Lewis, 1990: "Despite the easing of East-West tensions, the effectiveness of the United Nations as a forum for global agreement remains impaired by a widening North-South split between industrialized and developing nations."

With regard to Braudel's observation that "the desert lies in wait for arable land," it has been estimated that "four-fifths of the world's rangeland and three-quarters of its cropland have been transformed into desert since agriculture began" (Agnew & Warren, 1990: 15).

31. At the end of the thirteenth century the population of Paris was barely 60,000 (Russell, 1958: 61). At the end of the fourteenth century, after the Black Death had taken its toll, London was estimated to have been reduced to a population of 50,000 (Clapham, 1963: 78). By the middle of the fifteenth century it had recovered only to 75,000 (Chandler & Fox, 1974: 316).

32. Based on a count of entries in the *DSB*. See Figure 6.

33. Its northward irruption was blocked, in the West by the Franks and in the East by the Byzantines and by Khazaria, the medieval empire between the Black and Caspian Seas whose nobility had converted to Judaism in the eighth century.

CHAPTER FIVE. ABOVE THE FORTIETH PARALLEL

1. Cf. White (1962a), 76: "In the full sense of the vernacular, the Middle Ages, from the tenth century onward, were full of beans."

2. Hand mills for grinding grain are quite ancient and had in fact been carried about by Philip's Macedonian army (Donald Engels, 1978: 12).

3. Cf. Bender (1988), 5: "Most of the early universities were, like Bologna, located in medium-sized (though prosperous) cities, often, at least in Italy, dominated by a more powerful neighboring city."

4. See also Bender (1988), 3: "The revival of cities and the revival of learning under the aegis of the university in medieval Europe were coincident, and since then the city and the university have shared more of a common history than we usually recognize."

5. Cf. Haskins (1957), 12: "In northern Europe the origin of universities must be

sought at Paris, in the cathedral school of Notre-Dame. By the beginning of the twelfth century in France and the Low Countries learning was no longer confined to monasteries but had its most active centres in the schools attached to cathedrals."

6. For a more extensive discussion of some of these issues, see Sayili (1947–1948). See also Makdisi (1990).

7. Subjects like the sciences, history, and philosophy, which lend themselves to analytical treatment, form the core of the university curriculum. In contrast, the creative arts, which rest more heavily on imagination than on analysis, have never found a secure place in the university. Even engineering, which, when properly practiced, is highly intuitive, entered the university only with difficulty, and only after being remodelled as "applied science."

8. Joan of Arc (1412–1431), an illiterate girl of 17, was able to best the experienced English commanders partly because artillery was so new that experience carried little advantage, as is the case with any new technology that is not derived from accumulated knowledge. Joan's fellow commander, the Duke of Alençon, praised her military ability, "especially in the placing of artillery, for in that she acquitted herself magnificently" (quoted in Gies, 1981: 86). To appreciate the sweep of the "Joan of Arc syndrome," the reader need only reflect on the careers of Stephen Jobs and the other *Wunderkinder* of the early years of the computer revolution. Jobs, a college dropout, invented the *Apple I* in 1976 at the age of 21. Along with Stephen Wozniak, another dropout five years his senior, he designed the *Apple* in his bedroom and built the prototype in his garage. One of IBM's software programs was created by Raymond Kurzweil at age 12. Also, in the early history of radio and television, several of the key inventions were made by very young men: Paul Nipkow (1860–1940) invented the scanning disc and Edwin Armstrong (1890–1954) the feedback circuit and the superheterodyne circuit when they were in their early twenties. And Philo Farnsworth (1906–1971) invented the picture tube at the age of 16. (These data on the development of radio and television were called to my attention by one of my students, Paul J. Weihs.)

9. The Mons Meg, cast for the Burgundians in the 1440s, was 15 feet long, weighed 17,000 pounds, and threw a stone ball that was nearly two feet in diameter (G. Parker, 1988: 7–8).

10. G. Parker, 1988: 1; Roberts, 1967a: 203; McNeill, 1982: 111. McNeill suggests that the latter figure was probably closer to 300,000.

11. Cf. McNeill (1982), 109: "Small wonder, therefore, that even though he tripled Castile's taxes between 1556 and 1577, Philip II had to repudiate his debts on four separate occasions (1557, 1560, 1575, 1596) and never managed to pay his soldiers on time." See also Kennedy, 1987: 46–48. In England, "ship-money," a tax on real property levied in the 1630s to finance the royal fleet, was one of the grievances that led to the overthrow of the monarchy.

12. For an elaboration of the role of the new gunpowder technology in establishing European hegemony across the Old and New worlds, see Cipolla (1965). Europe was favored over the Mediterranean world in the development of the new ordnance because of its plentiful resources of iron ore, especially after the introduction of the cast-iron gun in the sixteenth century.

13. He also noted the effects of the military revolution in extending state con-

trol over nonmilitary professions: "Concern to sustain the Military Revolution also led the state to intervene more directly in education"; and: "The Military Revolution equally led to a greater state intervention in medicine" (Duffy, 1980: 5).

14. Just prior to the Black Death in the fourteenth century, the University of Padua had three professors—in medicine, surgery, and astrology (Campbell, 1931: 96).

15. Although Hermes' existence as a historical personality is doubtful, the legends surrounding his name are sufficiently significant to have earned him a "biography" in the DSB.

16. Cf. Hansen (1986), 128: "Because its aim was some useful accomplishment, not mere knowledge or understanding, magic was a practical technology rather than a science. . . . Even in cases where magic was employed to learn something, the goal was not philosophical understanding of the natures of things, but practical knowledge, such as the paternity of a child, the virginity of a woman, the location of a lost object, the success of an enterprise, or the prognosis of a disease. It is a mistake to view medieval magic as a philosophy, a source of knowledge, a way of thinking."

17. Brahe was occasionally assisted in his astronomical research by his sister, Sophia, who, according to Brahe's biographer, John Dreyer, was "highly educated, and not only conversant with classical literature, but also well acquainted with astrology and alchemy, and therefore in every way fit to assist her great brother" (Dreyer, 1963: 73).

18. Aficionados of the American system of federal research grants will nod knowingly. The practice is not confined to the United States: "Applying for research grants, as anyone who has ever served on a Grants Committee will be only too well aware, scientists are not unnaturally tempted to claim for their work that it will have, or at the very least might have, 'technological spin-offs'. This is so even when its sole interest is intellectual, its motivation that 'idle curiosity' which, we have recently been told in Australia, no government can be expected to support, however much some of us may think of it as the central core of human achievement" (Passmore, 1989: 567).

19. He also conceded that Newton's "secret alchemical investigations" present a problem (Thomas, 1971: 644).

20. From John Evelyn's Panegyric, addressed to the king (quoted in R. F. Jones, 1961: 179).

21. For the few instances where the Royal Society was drawn into public service during the eighteenth century, see J. E. McClellan, 1985: 29–34.

22. The philosophical use of "male" and "female" principles had Hermetic roots.

23. The shortage became even more severe in the first half of the eighteenth century, with one-third of the British merchant fleet being "colonial built," constructed in the American colonies where timber was plentiful (Clapham, 1963: 236).

24. Some writers have favored alternate explanations of the failure of the Baconian program of applied science at the Royal Society. G. N. Clark believed that it was the result of "public calamities"—the Dutch Wars and the fire of London

(Clark, 1937: 17). Margaret 'Espinasse contested Clark's opinion, suggesting that the "contempt for applied science" may have been a consequence of Newton's achievement, which distracted scientists from utilitarian research ('Espinasse, 1958: 84ff.). It may be noted that there was a decline in the number of scientists during the third quarter of the seventeenth century. (This observation is based on a count of entries in the *DSB*.)

25. The embarrassment was apparently felt by the scientists themselves. Although the Royal Society was circumvented by the Board of Longitude, the board was dominated by scientists, and there are indications that they would have preferred an astronomical solution to one achieved through the crafts. In the 1760s, even after Harrison's clock had been sucessfully tested in transatlantic trials, the board withheld the prize, hoping for solutions based on lunar observations. Harrison's entitlement was honored in 1773 only after he petitioned the House of Commons (Battison, 1972: 130).

26. "It is clear that all attempts at mathematically precise theories of ballistics or tables of range were irrelevant to the crude guns of the time. . . . Trial and error was a much better guide than mathematics. . . . Consequently sixteenth-century gunners from Spain and elsewhere fired at close range and without assistance from mathematics" (Goodman, 1988: 125).

27. The confusion is now petrified in dictionary definitions. Technology is defined as "the application of science, especially to industrial or commercial objectives"; and engineering as "the application of scientific principles to practical ends" (*American Heritage Dictionary of the English Language* [1971]).

28. In the eighteenth century the beliefs were propagated that the invention of the steam engine owed something to an alleged correspondence between Newcomen, the ironmonger, and Hooke, the scientist, and that Watt's separate condenser had been suggested by the principle of latent heat enunciated by Watt's friend, the chemist Joseph Black. Both allegations have been discredited (Jenkins, 1936–1937; Fleming, 1952; see also Dorn, 1976). Watt presented a convincing description of the events that led to the separate condenser, and he explicitly denied any dependence of his invention on Black's principle.

29. For the history of kinematics from the time of Watt, see Ferguson (1962). For thermodynamics, see Kuhn (1969).

30. Cultural and intellectual historians are particularly vulnerable to this misunderstanding of the engineering act. For a more knowing opinion by an economic historian and an engineer, see Nathan Rosenberg's and Walter Vincenti's collaborative study of British engineering and technical knowledge in the nineteenth century: "There is a widespread tendency to regard technology as involving essentially the application of knowledge derived from science. We believe that such a model of engineering activity is seriously deficient on a variety of grounds" (Rosenberg & Vincenti, 1978: 71).

31. Cf. Alter (1987), 71–72: "Even if it is not correct to say that the state gave no support for basic research at this time, it was in general granted hesitantly and unwillingly—according to *Nature*, with 'comparative indignity' and 'in a grudging way'—and was limited to an absolute minimum."

CHAPTER SIX. WESTERN SCIENCE GOES WEST

1. The Los Angeles Department of Water and Power has been said to rule Owens Valley "like a feudal barony" (Reinhold, 1989).

2. Open hostility to Mormonism survived through the first quarter of the twentieth century. The Saints' early advocacy of polygamy often served as a pretext to justify persecution. Some early silent films, such as *Marriage or Death* and *Deadwood Dick Spoils Brigham Young*, were sharply anti-Mormon.

3. On Mormon communitarianism prior to their arrival in Utah, see Arrington, 1953.

4. Mormons have continued to build temples all over the world, including one built in communist East Germany in 1985.

5. "A large share of the prominent administrators of Iran have been educated in Utah" (Arrington & Bitton, 1979: 316).

6. Despite lurid and exaggerated accounts of plural marriage among the early Mormons, the practice was not as exciting as its critics pictured it. Probably fewer than 2 percent of Mormon men lived in polygamy during the years that it was practiced (Widtsoe, 1987: 390).

7. At present, the federal government "owns most of the nation's aqueducts and other systems of transporting water" and "controls an estimated 25 percent of all water sources in the [American] West" (Shabecoff, 1988). It also owns 86 percent of the state of Nevada.

8. The escalation of the projected cost of the supercollider to eight billion dollars has sharpened the controversy over the rationale for the expenditure of public revenues for basic research. According to a recent news report: "Supporters of the accelerator have sometimes sought to sell the project to members of Congress on the ground that it may lead to important innovations in practical technology and would enhance scientific education at all levels. But most scientists agree that the real object in building the project is essentially intellectual and philosophical (some even say theological) rather than practical" (M. Browne, 1990: C6).

9. In 1954 the American physicist J. Robert Oppenheimer was deprived of his security clearance "mainly on grounds of 'defects of character'" (Peierls, 1974: 217). During the hearings that were held to judge him, Oppenheimer was questioned about the relative values he attached to loyalty to his country and to his friends.

10. Cf. Dickson (1988), 108–9: "The Reagan administration has also made it clear that the significant increases secured, with congressional blessing, in the basic research budget of other agencies—in particular the National Science Foundation—have been targeted in areas such as mathematics, physics, and engineering partly on the extent to which these subjects contribute toward strengthening the nation's military technology."

11. In opposition to funding the social sciences, one congressman warned, in somewhat dated phraseology, that patronizing the social sciences would result in "a lot of short-haired women and long-haired men messing into everybody's personal affairs" (quoted in Greenberg, 1967: 109).

12. Milton Lomask commented that the National Science Board had endorsed IRRPOS "with surprising enthusiasm in view of the propensity of many of its

members, including Chairman Handler, to flinch at the prospect of putting into practical research money they would rather have seen bestowed on investigations of a purer nature" (Lomask, 1975: 219).

13. Personal communication from Syl McNinch, former member of the NSF task force that developed the RANN concept.

14. The argument that physics is inherently more costly than anthropology is questionable. If the eight billion dollars being earmarked for the superconducting supercollider were made available to paleoanthropology the field would expand and there would be no shortage of sites at which anthropologists would enjoy digging. Museum basements are crammed with uncatalogued bones and artifacts that alone could occupy generations of happy anthropologists and archaeologists.

15. Market economies are becoming increasingly centralized under the pressure of intense international competition. The European Community recently set minimum prices for Japanese computer chips: "The action, which mirrors one taken by the United States more than three years ago, moves the world further toward a system in which prices of the crucial silicon circuits . . . are controlled by government decree rather than by pure market forces" (Pollack, 1990: D1).

REFERENCES

Adams, R. E. W.
1980   "Swamps, Canals, and the Locations of Ancient Maya Cities." *Antiquity* 54:206–14.
Adams, Robert McC.
1981   *Heartland of Cities*. Chicago: University of Chicago Press.
Agnew, Clive, and Andrew Warren
1990   "Sand Trap." *The Sciences* (March/April).
Alter, Peter
1987   *The Reluctant Patron*. Oxford: Berg Publishers.
Altman, Lawrence K.
1989   "Departing N.I.H. Chief Issues Warning on Politics." *New York Times*, June 20: C3.
Anderson, Perry
1974   *Lineages of the Absolutist State*. London: NLB.
Andrewes, Antony
1978   *The Greeks*. New York: Norton.
Aristophanes
1938   *The Clouds*. In Oates and O'Neill, 1938: vol. 2, 541–99.
Aristotle
1941   *Politics*. In McKeon, 1941: 1113–1316.
Ariyapala, M. B.
1956   *Society in Medieval Ceylon*. Colombo, Sri Lanka: Department of Cultural Affairs.
Armillas, Pedro
1971   "Gardens on Swamps." *Science* 174:653–61.
Arrington, Leonard J.
1953   "Early Mormon Communitarianism: The Law of Consecration and Stewardship." *Western Humanities Review* 7 (Autumn): 341–69.
1958   *Great Basin Kingdom*. Cambridge: Harvard University Press.
Arrington, Leonard J., and Davis Bitton
1979   *The Mormon Experience*. New York: Alfred A. Knopf.
Ashley, Maurice
1961   *England in the Seventeenth Century*. Baltimore: Penguin Books.
Athenaeus
1927   *The Deipnosophists*. Vol. 1. London: William Heinemann.
Aveni, Anthony F.
1980   *Skywatchers of Ancient Mexico*. Austin: University of Texas Press.

1982a   "Archaeoastronomy in the Maya Region: 1970–1980." In Aveni, 1982b: 1–30.

Aveni, Anthony F., ed.
1975   *Archaeoastronomy in Pre-Columbian America.* Austin: University of Texas Press.
1977   *Native American Astronomy.* Austin: University of Texas Press.
1982b   *Archaeoastronomy in the New World.* Cambridge: Cambridge University Press.

Avignon, Colette
1971   "John Evelyn." *DSB:* vol. 4, 494–97.

Bailes, Kendall E., ed.
1985   *Environmental History: Critical Issues in Comparative Perspective.* Lanham, Md.: University Press of America.

Bailey, Anne M., and Josep R. Llobera, eds.
1981   *The Asiatic Mode of Production.* London: Routledge & Kegan Paul.

Battison, Edwin A.
1972   "John Harrison." *DSB:* vol. 6, 130–31.

Beckinsale, R. P.
1975   "Friedrich Ratzel." *DSB:* vol. 11, 308–10.

Ben-David, Joseph
1971   *The Scientist's Role in Society.* Englewood Cliffs, N.J.: Prentice-Hall.

Bender, Thomas, ed.
1988   *The University and the City.* New York: Oxford University Press.

Beyerchen, Alan D.
1977   *Scientists under Hitler.* New Haven: Yale University Press.

Bielenstein, Hans
1950   "An Interpretation of the Portents in the Ts'ien-Han-Shu." *Bulletin of the Museum of Far Eastern Antiquities* (Stockholm) 22: 127–43.

Birch, Thomas
1756–1757   *The History of the Royal Society of London,* 4 vols. London.

Bivar, A. D. H.
1972   "Cavalry Equipment and Tactics on the Euphrates Frontier." *Dumbarton Oaks Papers,* no. 26: 271–91.

Bloch, Marc
1961   *Feudal Society.* 2 vols. Chicago: University of Chicago Press.

Boardman, John
1963   "Greek Archaeology on the Shores of the Black Sea." *Journal of Hellenic Studies* 83:34–51.

Bodde, Derk
1965   "Feudalism in China." In Coulborn, 1965: 49–92.

Bonelli, M. L. Righini, and William R. Shea, eds.
1975   *Reason, Experiment, and Mysticism in the Scientific Revolution.* New York: Science History Publications.

Boswall, Jeffery
1987   "Russia Is for the Birds." *Discover* 8 (March): 78–83.

Botsford, George Willis, and Charles Alexander Robinson, Jr.
1948   *Hellenic History.* New York: Macmillan.

Bowman, Kenneth P.
1988   "Global Trends in Total Ozone." *Science* 239:48–50.
Boyer, Carl B.
1959   *The History of the Calculus and Its Conceptual Development*. New York: Dover Publications.
Braudel, Fernand
1975   *The Mediterranean*. 2 vols. New York: Harper & Row.
Bray, Francesca
1986   *The Rice Economies*. Oxford: Basil Blackwell.
British Sessional Papers
1801   "Report from the Select Committee upon the Improvement of the Port of London." Vol. 3. London: House of Commons.
Brohier, R. L.
1975   *Food and the People*. Colombo, Sri Lanka: Lake House Investments.
Browne, E. G.
1983   *Arabian Medicine*. Westport, Conn.: Hyperion Press.
Browne, Malcolm
1990   "Supercollider's Rising Cost Provokes Opposition." *New York Times*, May 29: C1.
Bulmer-Thomas, Ivor
1971   "Euclid." *DSB*: vol. 4, 414–37.
Burke, John G.
1966   "Bursting Boilers and the Federal Power." *Technology and Culture* 7:1–23.
Burke, John G., ed.
1987   *Science and Culture in the Western Tradition*. Scottsdale, Ariz.: Gorsuch Scarisbrick.
Burn, A. R.
1968   *The Warring States of Greece*. New York: McGraw-Hill.
Burnet, John
1914   *Greek Philosophy: Thales to Plato*. London: Macmillan.
Bush, Vannevar
1945   *Science, the Endless Frontier*. Washington, D.C.: U.S. Government Printing Office.
Butzer, Karl
1976   *Early Hydraulic Civilization in Egypt*. Chicago: University of Chicago Press.
Cameron, Alan
1969   "The Last Days of the Academy at Athens." *Proceedings of the Cambridge Philological Society*, n.s. 15: 7–29.
Campbell, Anna Montgomery
1931   *The Black Death and Men of Learning*. New York: Columbia University Press.
Carnot, Sadi
1824   "Reflections on the Motive Power of Fire and on Machines Fitted to Develop That Power." In Mendoza, 1960: 1–59.

Carter, Vernon Gill, and Tom Dale
  1974   *Topsoil and Civilization.* Norman: University of Oklahoma Press.
Cary, Max
  1949   *The Geographic Background of Greek and Roman History.* Oxford: Oxford University Press.
Catto, J. I., ed.
  1984   *The History of the University of Oxford.* Vol. 1, *The Early Oxford Schools.* Oxford: Oxford University Press.
Chambers, Mortimer, Raymond Grew, David Herlihy, Theodore K. Rabb, and Isser Woloch
  1987   *The Western Experience.* 4th ed. Vol. 2. New York: Alfred A. Knopf.
Chandler, Tertius, and Gerald Fox
  1974   *3000 Years of Urban Growth.* New York: Academic Press.
Chen, Allan
  1987   "Unraveling Another Mayan Mystery." *Discover* 8 (June): 40–49.
Chi Ch'ao-ting
  1963   *Key Economic Areas in Chinese History As Revealed in the Development of Public Works for Water-Control.* 1936. New York: Paragon Book Reprint.
Childe, V. Gordon
  1942   *What Happened in History.* Harmondsworth, England: Penguin Books.
  1951   *Man Makes Himself.* New York: New American Library.
Cipolla, Carlo M.
  1965   *Guns, Sails, and Empires.* New York: Pantheon Books.
Cipolla, Carlo M., ed.
  1970   *The Economic Decline of Empires.* London: Methuen.
Clagett, Marshall
  1963   *Greek Science in Antiquity.* New York: Collier Books.
  1970   "Archimedes," *DSB:* vol. 1, 213–31.
  1989   *Ancient Egyptian Science.* Vol. 1 (2 tomes). Philadelphia: American Philosophical Society.
Clagett, Marshall, ed.
  1969   *Critical Problems in the History of Science.* Madison: University of Wisconsin Press.
Clapham, John
  1963   *A Concise Economic History of Britain: From the Earliest Times to 1750.* Cambridge: Cambridge University Press.
Clark, G. N.
  1937   *Science and Social Welfare in the Age of Newton.* Oxford: Oxford University Press.
Closs, Michael P.
  1986a   "The Mathematical Notation of the Ancient Maya." In Closs, 1986b: 291–369.
  1986b   *Native American Mathematics.* Austin: University of Texas Press.
Clow, Archibald, and Nan L. Clow
  1956   "The Timber Famine and the Development of Technology." *Annals of Science* 12:85–102.

Coe, Michael D.
1964   "The Chinampas of Mexico." *Scientific American* 211 (July): 90–98.
1975   "Native Astronomy in Mesoamerica." In Aveni, 1975: 3–31.
1986   Preface to Schele and Miller, 1986.
Cohen, G. A.
1978   *Karl Marx's Theory of History: A Defense.* Princeton: Princeton University Press.
Cohen, Mark Nathan
1977   *The Food Crisis in Prehistory.* New Haven: Yale University Press.
Cohen, Ronald, and Elman R. Service, eds.
1978   *Origins of the State.* Philadelphia: Institute for the Study of Human Issues.
Collier, Donald
1955   "Development of Civilization on the Coast of Peru." In Steward et al., 1955: 19–27.
Conant, James Bryant, ed.
1957a   "The Overthrow of the Phlogiston Theory." In Conant, 1957b: 65–115.
1957b   *Harvard Case Histories in Experimental Science.* Vol. 1. Cambridge: Harvard University Press.
Congressional Research Service (Library of Congress)
1986   *The Nobel-Prize Awards in Science as a Measure of National Strength in Science.* Washington, D.C.: U.S. Government Printing Office.
Conzen, Michael P.
1987   "Land and Society through the Ages: A Cartographic Essay." In Chambers et al., 1987.
Coulborn, Rushton, ed.
1965 [1956]   *Feudalism in History.* Hamden, Conn.: Archon Books.
Critchley, Macdonald
1987   "Really the Blues." *New York Review of Books* 34, no. 2 (February 12): 5–8.
Crombie, A. C.
1959   *Medieval and Early Modern Science.* 2 vols. New York: Doubleday Anchor.
1969   "The Significance of Medieval Discussions of Scientific Method for the Scientific Revolution." In Clagett, 1969: 79–101.
Crombie, A. C., ed.
1963   *Scientific Change.* New York: Basic Books.
Crombie, A. C., and J. D. North
1970   "Roger Bacon." *DSB:* vol. 1, 377–85.
Crosby, Alfred W.
1986   *Ecological Imperialism: The Biological Expansion of Europe, 900–1900.* Cambridge: Cambridge University Press.
Curry, Patrick
1987a   "Saving Astrology in Restoration England: 'Whig' and 'Tory' Reforms." In Curry, 1987b: 245–59.
Curry, Patrick, ed.
1987b   *Astrology, Science and Society.* Woodbridge, England: Boydell Press.

Daniel, Glynn
1980   "Megalithic Monuments." *Scientific American* 243 (July): 78–90.
Danilova, L. V.
1971   "Controversial Problems of the Theory of Precapitalist Societies."
        Soviet Anthropology and Archeology 9:269–328.
Dannenfeldt, Karl H.
1972   "Hermes Trismegistus." *DSB:* vol. 6, 305–6.
Diamond, Jared
1987   "The Worst Mistake in the History of the Human Race." *Discover* 8
        (May): 64–66.
Dicks, D. R.
1959   "Thales." *The Classical Quarterly* 53 (n.s. 9): 294–309.
Dickson, David
1988   *The New Politics of Science.* Chicago: University of Chicago Press.
Dobbs, Betty Jo Teeter
1975   *The Foundations of Newton's Alchemy, or "The Hunting of the
        Greene Lyon."* Cambridge: Cambridge University Press.
Dorn, Harold
1970   *The Art of Building and the Science of Mechanics.* Ph.D. diss., Prince-
        ton University.
1974   "Thomas Newcomen." *DSB:* vol. 10, 36–37.
1976   "James Watt." *DSB:* vol. 14, 196–99.
1979   "Hugh Lincoln Cooper and the First Détente." *Technology and Cul-
        ture* 20:322–47.
Doyel, David E.
1979   "The Prehistoric Hohokam of the Arizona Desert." *American Scien-
        tist* 67:544–54.
Doyen-Higuet, Anne-Marie
1985   "The *Hippiatrica* and Byzantine Veterinary Medicine." In Scar-
        borough, 1985: 111–20.
Dreyer, J. L. E.
1963   *Tycho Brahe.* New York: Dover Publications.
DSB   *See* Gillispie, 1970–1980.
Duffy, Michael
1980   *The Military Revolution and the State 1500–1800.* Exeter: University
        of Exeter.
Dunn, Stephen P.
1982   *The Fall and Rise of the Asiatic Mode of Production.* London: Rout-
        ledge & Kegan Paul.
Eberhard, Wolfram
1957   "The Political Function of Astronomy and Astronomers in Han
        China." In Fairbank, 1957: 33–70.
1965   *Conquerors and Rulers: Social Forces in Medieval China.* Leiden,
        The Netherlands: E. J. Brill.
Edelstein, Ludwig
1957   "Recent Trends in the Interpretation of Ancient Science." In Wiener
        and Noland, 1957: 90–121.

1963　"Motives and Incentives for Science in Antiquity." In Crombie, 1963: 15–41.

Elliott, John
1987　"India's Reluctant Finance Minister." *Financial Times* (London), August 26: 15.

Ellis, Florence Hawley
1975　"A Thousand Years of the Pueblo Sun-Moon-Star Calendar." In Aveni, 1975: 59–87.

Emlyn-Jones, C. J.
1980　*The Ionians and Hellenism.* London: Routledge & Kegan Paul.

Engelbert, Ernest A.
1984　*Water Scarcity.* Berkeley: University of California Press.

Engels, Donald W.
1978　*Alexander the Great and the Logistics of the Macedonian Army.* Berkeley: University of California Press.

Engels, Friedrich
1939　*Herr Eugen Duhring's Revolution in Science.* New York: International Publishers.

England, J. Merton
1982　*A Patron for Pure Science.* Washington, D.C.: National Science Foundation.

'Espinasse, Margaret
1958　"The Decline and Fall of Restoration Science." *Past and Present* 14 (November): 71–89.

Fairbank, John K., ed.
1957　*Chinese Thought and Institutions.* Chicago: University of Chicago Press.

Farmer, B. H.
1957　*Pioneer Peasant Colonization in Ceylon.* London: Oxford University Press.

Farrington, Benjamin
1980　*Greek Science.* Nottingham: Spokesman.

Ferguson, Eugene S.
1962　"Kinematics of Mechanisms from the Time of Watt." *Contributions from the Museum of History and Technology.* U.S. National Museum Bulletin 228, no. 27.

Fernando, A. Denis N.
1980　"Major Ancient Irrigation Works of Sri Lanka." *Journal of the Sri Lanka Branch of the Royal Asiatic Society* 22 (n.s., special no.).

1982　"The Ancient Hydraulic Civilization of Sri Lanka in Relation to its Natural Resources." *Journal of the Sri Lanka Branch of the Royal Asiatic Society* 27 (n.s., special no.).

Ferruolo, Stephen C.
1988　"*Parisius-Paradisus:* The City, Its Schools, and the Origins of the University of Paris." In Bender, 1988: 22–43.

Fischer, David Hackett
1981   "Climate and History: Priorities for Research." In Rotberg and Rabb, 1981: 241–50.
Fleming, Donald
1952   "Latent Heat and the Invention of the Watt Engine." *Isis* 43:3–5.
Fogel, Joshua A.
1988   "The Debates over the Asiatic Mode of Production in Soviet Russia, China, and Japan." *American Historical Review* 93, no. 1 (February): 56–79.
Framji, K. K., and I. K. Mahajan
1969   *Irrigation and Drainage in the World.* Vol. 1. New Delhi: International Commission on Irrigation and Drainage.
Frazier, Kendrick
1980   "The Anasazi Sun Dagger." *Science 80,* no. 1: 56–67.
Gaile, Gary L., and Cory J. Willmott
1989   *Geography in America.* Columbus, Ohio: Merrill.
Gardiner, Alan H.
1947   *Ancient Egyptian Onomastica.* Vol. 1. Oxford: Oxford University Press.
Gascoigne, Robert Mortimer
1984   *A Historical Catalogue of Scientists and Scientific Books.* New York: Garland.
Genovese, Eugene D., and Leonard Hochberg
1989   *Geographic Perspectives in History.* Oxford: Basil Blackwell.
Georgi, Howard, Sheldon L. Glashow, and Kenneth D. Lane
1987   "Waiting for the Godot of Superconductors." *New York Times,* May 14: A30.
Ghalioungui, Paul
1983   *The Physicians of Pharaonic Egypt.* Mainz am Rhein, Germany: Verlag Philipp von Zabern.
Gies, Frances
1981   *Joan of Arc.* New York: Harper & Row.
Gillings, R. J.
1978   "The Mathematics of Ancient Egypt." *DSB:* vol. 15, 681–705.
Gillispie, Charles Coulston, ed.
1970–1980   *Dictionary of Scientific Biography.* 16 vols. New York: Charles Scribner's Sons.
Gillispie, Charles Coulston
1980   *Science and Polity in France at the End of the Old Regime.* Princeton: Princeton University Press.
Gingerich, Owen
1973   "Johannes Kepler." *DSB:* vol. 7, 289–312.
Gladwin, Harold S., Emil W. Haury, Edwin B. Sayles, and Nora Gladwin
1938   *Excavations at Snaketown, Material Culture.* Medallion Papers, no. 25. Globe, Ariz.: Gila Pueblo.
Godes, M.
1981   "The Reaffirmation of Unilinealism." In Bailey and Llobera, 1981: 99–105.

Goodman, David
1988 *Power and Penury.* Cambridge: Cambridge University Press.
Gorbachev, Mikhail
1987 *Perestroika.* New York: Harper & Row.
Gould, Stephen Jay
1988 "To Him, Fossils Said the Darnedest Things." *New York Times,* February 14: sec. 7, 14–15.
Graham, Loren R.
1989 "Russian and Soviet Science and Technology." *History of Science Society Newsletter* 18, no. 4 (Supplement): 1–8.
Grant, Michael
1982 *From Alexander to Cleopatra.* New York: Charles Scribner's Sons.
Greenberg, Daniel S.
1967 *The Politics of Pure Science.* New York: New American Library.
Gunther, R. T.
1931 *Early Science in Oxford.* Vol. 8. Oxford.
1935 *Early Science in Oxford.* Vol. 10. Oxford.
Hall, A. E., G. H. Cannel, and H. W. Lawton, eds.
1979 *Agriculture in Semi-Arid Environments.* Berlin: Springer-Verlag.
Halsell, Grace
1990 "Princes of Learning." *Aramco World* 41, no. 1 (January–February): 12–13.
Hamarneh, Sami
1962 "Development of Hospitals in Islam." *Journal of the History of Medicine and Allied Sciences* 17:366–84.
Hamilton, S. B.
1958 "Building and Civil Engineering Construction." In Singer et al., 1954–1958: vol. 4, 442–88.
Hansen, Bert
1986 "The Complementarity of Science and Magic before the Scientific Revolution." *American Scientist* 74, no. 2 (March–April): 128–36.
Hardy, Kenneth R.
1974 "Social Origins of American Scientists and Scholars." *Science* 185:497–506.
Harris, David R.
1978 "The Agricultural Foundations of Lowland Maya Civilization: A Critique." In Harrison and Turner, 1978: 301–23.
Harris, Marvin
1977 *Cannibals and Kings.* New York: Vintage Books.
Harrison, Peter D., and Billie Lee Turner II, eds.
1978 *Pre-Hispanic Maya Agriculture.* Albuquerque: University of New Mexico Press.
Hartshorne, Richard
1939 *The Nature of Geography.* Lancaster, Pa.: Association of American Geographers.
Haskins, Charles Homer
1957 *The Rise of the Universities.* Ithaca: Cornell University Press.

Hau, Friedrun R.
1979   "Gondeschapur—eine Medizinschule aus dem 6. Jahrhundert n. Chr." *Gesnerus* 36:98–115.
Haury, Emil W.
1976   *The Hohokam: Desert Farmers and Craftsmen.* Tucson: University of Arizona Press.
Heath, Thomas
1921   *A History of Greek Mathematics.* Vol. 1. Oxford: Oxford University Press.
Heinerman, John, and Anson Shupe
1985   *The Mormon Corporate Empire.* Boston: Beacon Press.
Herodotus
1910   *The History of Herodotus.* 2 vols. Translated by George Rawlinson. London: E. P. Dutton.
Hessen, Boris
1971   "The Social and Economic Roots of Newton's 'Principia'." In *Science at the Crossroads,* 1971: 147–212.
Hindess, Barry, and Paul Q. Hirst
1975   *Pre-capitalist Modes of Production.* London: Routledge & Kegan Paul.
Hitti, Philip K.
1961   *The Near East in History.* Princeton: Van Nostrand.
Hoberman, Barry
1987   "From Cuneiform to Topkapi." *Aramco World* 38, no. 2 (March–April): 2–5.
Hobsbawm, Eric J., ed.
1964   *Pre-Capitalist Economic Formations.* New York: International Publishers.
Holt, P. M., Ann K. S. Lambton, and Bernard Lewis, eds.
1970   *The Cambridge History of Islam.* Vol. 2. Cambridge: Cambridge University Press.
Hooke, Robert
1661   *An Attempt for the Explication of the Phaenomena.* In Gunther, 1935: 1–50.
1678   *Lectures De Potentia Restitutiva, or Of Spring Explaining the Power of Springing Bodies.* In Gunther, 1931: 331–56.
Hopwood, Derek, ed.
1990   *Studies in Arab History.* New York: St. Martin's Press.
Hunt, Eva
1972   "Irrigation and the Socio-political Organization of Cuicatec Cacicazgos." In F. Johnson, 1972: 162–259.
Hunt, Eva, and Robert C. Hunt
1978   "Irrigation, Conflict and Politics: A Mexican Case." In Cohen and Service, 1978: 69–123.
Hunter, James M.
1983   *Perspective on Ratzel's Political Geography.* Lanham, Md.: University Press of America.

Huntington, Ellsworth
1922   *Civilization and Climate.* New Haven: Yale University Press.
Huxley, G. L.
1966   *The Early Ionians.* London: Faber & Faber.
Hyams, Edward
1976   *Soil and Civilization.* New York: Harper & Row.
Hyde, J. K.
1988   "Universities and Cities in Medieval Italy." In Bender, 1988: 13–21.
Jacob, Margaret
1988   *The Cultural Meaning of the Scientific Revolution.* New York: Alfred
        A. Knopf.
Jacobsen, Thorkild, and Robert M. Adams
1958   "Salt and Silt in Ancient Mesopotamian Agriculture." *Science* 128:
        1251–58.
Jenkins, Rhys
1936–1937 "The Heat Engine Idea in the Seventeenth Century." *Transac-
        tions of the Newcomen Society* 17:1–11.
Johnson, Frederick, ed.
1972   *The Prehistory of the Tehuacan Valley.* Vol. 4: *Chronology and Irriga-
        tion.* Austin: University of Texas Press.
Johnson, George
1988   "The Scientist as Hedonist." *New York Times,* March 6: sec. 7, 7.
Jones, A. H. M.
1964   *The Later Roman Empire, 284–602.* 2 vols. Norman: University of
        Oklahoma. Reprint. Baltimore: Johns Hopkins University Press, 1986.
Jones, Richard Foster
1961   *Ancients and Moderns.* Berkeley: University of California Press.
Jungk, Robert
1958   *Brighter than a Thousand Suns.* New York: Harcourt Brace
        Jovanovich.
Katz, Friedrich
1972   *The Ancient American Civilizations.* New York: Praeger.
Keddie, Nikki R.
1981   "Socioeconomic Change in the Middle East since 1800: A Compara-
        tive Analysis." In Udovitch, 1981: 761–83.
Kennedy, Paul
1985   "The Reason Why." *New York Review of Books* 32, no. 4 (March 14):
        32–33.
1987   *The Rise and Fall of the Great Powers.* New York: Random House.
Khazanov, A. M.
1984   *Nomads and the Outside World.* Cambridge: Cambridge University
        Press.
Kibre, Pearl, and Nancy G. Siraisi
1978   "The Institutional Setting: The Universities." In Lindberg, 1978:
        120–44.
Kirby, Richard Shelton, Sidney Withington, Arthur Burr Darling, and Frederick
        Gridley Kilgour
1956   *Engineering in History.* New York: McGraw-Hill.

Kitto, H. D. F.
1957   *The Greeks.* Baltimore: Penguin Books.
Krader, Lawrence
1975   *The Asiatic Mode of Production.* Assen, The Netherlands: van
        Gorcum.
Kramer, Edna E.
1972   "Hypatia." *DSB:* vol. 6, 615–16.
Kramer, Samuel Noah
1959   *History Begins at Sumer.* New York: Doubleday Anchor Books.
Kuhn, Thomas S.
1969   "Energy Conservation as an Example of Simultaneous Discovery." In
        Clagett, 1969: 321–56.
1972   "Scientific Growth: Reflections on Ben-David's 'Scientific Role'."
        Minerva 10:166–78.
Lapidus, Ira M.
1981   "Arab Settlement and Economic Development of Iraq and Iran in the
        Age of the Umayyad and Early Abbasid Caliphs." In Udovitch, 1981:
        177–208.
Lattimore, Owen
1951   *Inner Asian Frontiers of China.* New York: American Geographical
        Society.
1962a   "The Frontier in History." In Lattimore, 1962b: 469–91.
1962b   *Studies in Frontier History.* London: Oxford University Press.
Lawton, H. W., and P. J. Wilke
1979   "Ancient Agricultural Systems in Dry Regions." In Hall, Cannel, and
        Lawton, 1979: 1–44.
Layton, Edwin
1971   "Mirror-Image Twins: The Communities of Science and Technology
        in Nineteenth-Century America." *Technology and Culture* 12:562–
        80.
Leach, E. R.
1959   "Hydraulic Society in Ceylon." *Past and Present,* no. 15 (April): 2–26.
Le Roy Ladurie, Emmanuel
1971   *Times of Feast, Times of Famine: A History of Climate since the Year
        1000.* New York: Doubleday.
Levenson, Joseph R.
1956   Review of Coulborn, 1965. *Far Eastern Quarterly* 15:569–72.
Lewis, Paul
1990   "Amid Jealousies, a Suspicion That the U.N. Works." *New York
        Times,* January 15: A6.
Lichtheim, George
1963   "Marx and the 'Asiatic Mode of Production'." *St. Antony's Papers,* no.
        14: 86–112.
Lindberg, David C., ed.
1978   *Science in the Middle Ages.* Chicago: University of Chicago Press.
Lloyd, G. E. R.
1970   *Early Greek Science: Thales to Aristotle.* New York: W. W. Norton.

1979   *Magic, Reason and Experience.* Cambridge: Cambridge University
         Press.
Lomask, Milton
1975   *A Minor Miracle.* Washington, D.C.: National Science Foundation.
Longrigg, James
1970   "Anaxagoras." *DSB:* vol. 1, 149–50.
1976   "Thales." *DSB:* vol. 13, 295–98.
Lounsbury, Floyd G.
1978   "Maya Numeration, Computation, and Calendrical Astronomy."
         DSB: vol. 15, 759–818.
Makdisi, George
1990   "On the Origin and Development of the College in Islam and the
         West." In Hopwood, 1990: 1–19.
Mao Tse-tung
1954   *Selected Works.* Vol. 3. New York: International Publishers.
Marshack, Alexander
1964   "Lunar Notation on Upper Paleolithic Remains." *Science* 146 (6
         Nov.): 743–45.
Marx, Karl
1936   *Capital.* New York: Random House.
1977   Preface to "A Critique of Political Economy." In McLellan, 1977: 388–
         92.
Marx, Karl, and Friedrich Engels
[1980]   *On Colonialism.* (Moscow: Foreign Languages Publishing House,
         n.d.) London: Lawrence & Wishart.
Matheny, Ray T.
1976   "Maya Lowland Hydraulic Systems." *Science* 193:639–46.
1978   "Northern Maya Lowland Water-Control Systems." In Harrison and
         Turner, 1978: 185–210.
McClellan, James E., III
1985   *Science Reorganized.* New York: Columbia University Press.
McEvedy, Colin, and Richard Jones
1978   *Atlas of World Population History.* Harmondsworth, England:
         Penguin.
McKeon, Richard, ed.
1941   *The Basic Works of Aristotle.* New York: Random House.
McLellan, David, ed.
1977   *Karl Marx: Selected Writings.* Oxford: Oxford University Press.
McNeill, William H.
1963   *The Rise of the West.* Chicago: University of Chicago Press.
1976   *Plagues and Peoples.* New York: Doubleday.
1982   *The Pursuit of Power.* Chicago: University of Chicago Press.
1983   *The Great Frontier.* Princeton: Princeton University Press.
Meinig, D. W.
1986   *The Shaping of America: A Geographical Perspective on 500 Years of
         History.* Vol. 1, *Atlantic America, 1492–1800.* New Haven: Yale Uni-
         versity Press.

206   REFERENCES

Mendelssohn, Kurt
  1970   "A Scientist Looks at the Pyramids." *American Scientist* 59:210–20.
Mendoza, E., ed.
  1960   *Reflections on the Motive Power of Fire.* New York: Dover Publications.
Merton, Robert K.
  1939   "Science and the Economy of Seventeenth-Century England." *Science and Society* 3:3–27.
  1970   *Science, Technology and Society in Seventeenth-Century England.* New York: Harper & Row.
Mill, John Stuart
  1961   *Principles of Political Economy.* New York: Augustus M. Kelley.
Miller, Timothy
  1985   *The Birth of the Hospital in the Byzantine Empire.* Baltimore: Johns Hopkins Press.
Mitchell, William P.
  1973   "The Hydraulic Hypothesis: A Reappraisal." *Current Anthropology* 14:532–34.
Morley, Sylvanus Griswold
  1946   *The Ancient Maya.* Stanford: Stanford University Press.
Naipaul, V. S.
  1981   *Among the Believers.* New York: Alfred A. Knopf.
Nakayama, Shigeru, and Nathan Sivin, eds.
  1973   *Chinese Science.* Cambridge: MIT Press.
Nasr, Seyyed Hossein
  1968   *Science and Civilization in Islam.* New York: New American Library.
Nawwab, Ismail I., Peter C. Speers, and Paul F. Hoye, eds.
  1980   *ARAMCO and Its World.* Washington, D.C.: Arabian American Oil Co.
Needham, Joseph
  1954   *Science and Civilisation in China.* Vol. 1. Cambridge: Cambridge University Press.
  1959   *Science and Civilisation in China.* Vol. 3. Cambridge: Cambridge University Press.
  1963   "Poverties and Triumphs of the Chinese Scientific Tradition." In Crombie, 1963: 117–53.
  1969a   "Science and Society in East and West." In Needham, 1969b: 190–217.
  1969b   *The Grand Titration.* Toronto: University of Toronto Press.
  1971a   *Science and Civilisation in China.* Vol. 4, part 3. Cambridge: Cambridge University Press.
  1971b   Foreword to *Science at the Crossroads,* 1971.
  1973   "The Historian of Science as Ecumenical Man: A Meditation in the Shingon Temple of Kongōsammai-in . . . on Kōyasan." In Nakayama and Sivin, 1973: 1–8.
  1984   *Science and Civilisation in China.* Vol. 6, part 2. Cambridge: Cambridge University Press.

Needham, Joseph, and Lu Gwei-Djen
1985   *Trans-Pacific Echoes and Resonances; Listening Once Again.* Singapore: World Scientific.
Neugebauer, Otto
1952   "Tamil Astronomy." *Osiris* 10:252–76.
1957   *The Exact Sciences in Antiquity.* Providence: Brown University Press.
Oates, Whitney J., and Eugene O'Neill, Jr.
1938   *The Complete Greek Drama.* 2 vols. New York: Random House.
O'Leary, Brendan
1989   *The Asiatic Mode of Production.* Oxford: Basil Blackwell.
Olson, Richard
1982   *Science Deified and Science Defied.* Berkeley: University of California Press.
Omohundro, John T.
1987   "The Folk Art of the Raised Bed." *Gardening* 11, no. 3 (May–June): 10–32.
Oppenheim, Leo A.
1978   "Man and Nature in Mesopotamian Civilization." *DSB:* vol. 15, 634–66.
Osborne, Robin
1987   *Classical Landscape with Figures.* London: George Philip.
Owen, G. E. L.
1970   "Aristotle." *DSB:* vol. 1, 250–58.
Palerm, Angel
1955   "The Agricultural Basis of Urban Civilization in Mesoamerica." In Steward et al., 1955: 28–42.
Parker, Geoffrey
1988   *The Military Revolution.* Cambridge: Cambridge University Press.
Parker, Richard A.
1978   "Egyptian Astronomy, Astrology, and Calendrical Reckoning." *DSB:* vol. 15, 706–27.
Passmore, John
1989   "Hearing Voices." *Times Literary Supplement* (May 26–June 1): 567.
Pasternak, Burton
1972   "The Sociology of Irrigation: Two Taiwanese Villages." In Willmot, 1972: 193–213.
Paz, Octavio
1987   "Food of the Gods." *New York Review of Books* 34, no. 3 (February 26): 3–7.
Pečírka, Jan
1964   "Die sowjetischen Diskussionen über die asiatische Produktionsweise und über die Sklavenhalterformation." *Eirene* 3:147–69.
1967   "Von der asiatischen Produktionsweise zu einer Marxistischen Analyse der frühen Klassengesellschaften." *Eirene* 6:141–74.
Peierls, Rudolf
1974   "J. Robert Oppenheimer." *DSB:* vol. 10, 213–18.

Petty, William
  1674   *The Discourse made before the Royal Society the 26 of November
          1674. Concerning the Use of Duplicate Proportion in sundry impor-
          tant Particulars: together with a New Hypothesis of Springing or
          Elastique Motions.* London.
Pirenne, Henri
  1937   *Economic and Social History of Medieval Europe.* New York: Har-
          court Brace Jovanovich.
Plato
  1892   *The Dialogues of Plato.* 3rd ed. Translated by B. Jowett. Vol. 1. Oxford:
          Clarendon Press.
Pollack, Andrew
  1990   "Europe Sets Prices for Japan's Chips." *New York Times,* January 23:
          D1.
Price, Barbara J.
  1978   "Secondary State Formation: An Explanatory Model." In Cohen and
          Service, 1978: 161–86.
Price, Derek de Solla
  1975   *Science since Babylon.* New Haven: Yale University Press.
Purver, Margery
  1967   *The Royal Society: Concept and Creation.* Cambridge: MIT Press.
Rabie, Hassanein
  1981   "Some Technical Aspects of Agriculture in Medieval Egypt." In
          Udovitch, 1981: 59–90.
Rashdall, Hastings
  1936   *The Universities of Europe in the Middle Ages.* 3 vols. Oxford: Oxford
          University Press.
Reagan, Michael D.
  1967   "Basic and Applied Research: A Meaningful Distinction?" *Science*
          155:1383–86.
Reinhold, Robert
  1989   "Dusty Harvest Is Near After Long Water Wars." *New York Times,*
          September 29: A10.
Renfrew, Colin
  1989   "The Origins of Indo-European Languages." *Scientific American* 261
          (October).
Roberts, Michael
  1967a   "The Military Revolution, 1560–1660." In Roberts, 1967b: 195–235.
  1967b   *Essays in Swedish History.* London: Weidenfeld and Nicolson.
Roebuck, Carl
  1959   *Ionian Trade and Colonization.* New York: Archaeological Institute
          of America.
Ronan, Colin A.
  1978   *The Shorter Science and Civilisation in China.* Vol. 1. Cambridge:
          Cambridge University Press.

Rosenberg, Nathan, and Walter Vincenti
1978   *The Britannia Bridge: The Generation and Diffusion of Technological Knowledge.* Cambridge: MIT Press.
Rossi, Paolo
1968   *Francis Bacon: From Magic to Science.* Chicago: University of Chicago Press.
Rostovtzeff, M.
1941   *The Social and Economic History of the Hellenistic World.* 3 vols. Oxford: Oxford University Press.
1957   *The Social and Economic History of the Roman Empire.* 2 vols. Oxford: Oxford University Press.
Rotberg, Robert I., and Theodore K. Rabb
1981   *Climate and History.* Princeton: Princeton University Press.
Rothblatt, Sheldon
1988   "Oxford University's Long Haul." *London Review of Books* 10, no. 2 (January 21): 12–15.
Rouse, Hunter, and Simon Ince
1957   *History of Hydraulics.* New York: Dover Publications.
Rowe, William T.
1985   "Approaches to Modern Chinese Social History." In Zunz, 1985: 236–96.
Runciman, Steven
1961   *Byzantine Civilization.* London: Methuen.
Russell, J. C.
1958   *Late Ancient and Medieval Population.* Philadelphia: American Philosophical Society.
Salaman, Redcliffe N.
1985   *The History and Social Influence of the Potato.* Cambridge: Cambridge University Press.
Saliba, George
1987   "The Role of Maragha in the Development of Islamic Astronomy: A Scientific Revolution before the Renaissance." *Revue de Synthèse* 108:361–73.
Sanger, David E.
1987   "U.S. Is Reviving Its Push to Build Fast Computers." *New York Times,* August 13: A1.
Sarton, George
1927   *Introduction to the History of Science.* Vol. 1. Baltimore: Williams & Wilkins.
1952   *A History of Science: Ancient Science through the Golden Age of Greece.* Cambridge: Harvard University Press.
1959   *A History of Science: Hellenistic Science and Culture in the Last Three Centuries B.C.* Cambridge: Harvard University Press.
Sayili, Aydin
1947–1948   "Higher Education in Medieval Islam." *Annales de l'université d'Ankara* 2:30–71.

1981　　The Observatory in Islam. New York: Arno Press.

Scarborough, John, ed.

1985　　"Symposium on Byzantine Medicine." Dumbarton Oaks Papers, no. 38.

Ščerbakiwskyj, V.

1926　　"Eine palaolithische Station in Honce (Ukraina)." Die Eiszeit 3–4:106–16.

Schaffer, Simon

1987　　"Newton's Comets and the Transformation of Astrology." In Curry, 1987b: 219–43.

Schele, Linda, and Mary Ellen Miller

1986　　The Blood of Kings. New York: George Braziller.

Schwartz, Benjamin

1953　　"A Marxist Controversy on China." Journal of Asian Studies 13:143–53.

Science at the Crossroads

1971　　London: Frank Cass.

Seidenberg, A.

1986　　"The Zero in Mayan Numerical Notation." In Closs, 1986b: 371–86.

Semple, Ellen Churchill

1911　　Influences of Geographic Environment on the Basis of Ratzel's System of Anthropo-Geography. New York: Henry Holt.

1931　　The Geography of the Mediterranean Region: Its Relation to Ancient History. New York: Henry Holt.

Service, Elman R.

1978　　"Classical and Modern Theories of the Origins of Government." In Cohen and Service, 1978: 21–34.

Shabecoff, Philip

1986　　"Aide Sees Need to Head Off Global Warming." New York Times, June 12: B8.

1989　　"Arid West Getting U.S. Water Broker." New York Times, January 1: 17.

Shepherd, William R.

1929　　Historical Atlas. New York: Henry Holt.

Sills, David L., ed.

1968　　International Encyclopedia of the Social Sciences. Vol. 7. New York: Macmillan.

Simon, Joan

1962　　"Stages in Social Development." Marxism Today 6 (June): 183–88.

Singer, Charles, E. J. Holmyard, and A. R. Hall, eds.

1954–1958　A History of Technology. 5 vols. New York: Oxford University Press.

Singleton, Charles S., ed.

1967　　Art, Science, and History in the Renaissance. Baltimore: Johns Hopkins Press.

Sivin, Nathan

1969　　"Cosmos and Computation in Early Chinese Mathematical Astronomy." T'oung Pao 55:1–73.

Skalník, Petr, and Timoteus Pokora
1966   "Beginning of the Discussion about the Asiatic Mode of Production in the U.S.S.R. and the People's Republic of China." *Eirene* 5:179–87.

Smith, Adam
1937   *An Inquiry into the Nature and Causes of the Wealth of Nations.* New York: Random House.

Sofaer, Anna, R. M. Sinclair, and L. E. Doggett
1982   "Lunar Markings on Fajada Butte, Chaco Canyon, New Mexico." In Aveni, 1982b: 169–81.

Sofaer, Anna, Volker Zinser, and Rolf M. Sinclair
1979   "A Unique Solar Marking Construct." *Science* 206:283–91.

Spate, O. H. K.
1968   "Ellsworth Huntington." In Sills, 1968: 26–27.

Stahl, William H.
1962   *Roman Science.* Madison: University of Wisconsin Press.

[Stalin, Joseph]
1939   "Dialectical and Historical Materialism." In *History of the Communist Party of the Soviet Union (Bolsheviks),* 105–31. New York: International Publishers.

Stencel, Robert, Fred Gifford, and Eleanor Moron
1976   "Astronomy and Cosmology at Angkor Wat." *Science* 193:281–87.

Stevens, William K.
1986   "Prehistoric Society: A New Picture Emerges." *New York Times,* December 16: C1.

1988   "Scientists Revive a Lost Secret of Farming." *New York Times,* November 22.

Steward, Julian H., Robert M. Adams, Donald Collier, Angel Palerm, Karl A. Wittfogel, and Ralph L. Beals
1955   *Irrigation Civilizations: A Comparative Study.* Washington, D.C.: Pan American Union.

Strabo
1959   *The Geography of Strabo.* Vol. 8. London: William Heinemann.

Stroup, Alice
1987   *Royal Funding of the Parisian Académie Royale des Sciences during the 1690s.* Philadelphia: American Philosophical Society.

Tarn, W. W.
1975   *Hellenistic Civilization.* New York: New American Library.

Taton, René, ed.
1963   *Ancient and Medieval Science.* Vol. 1. New York: Basic Books.

1971a   "Gian Domenico Cassini (Cassini I)." *DSB:* vol. 3, 100–104.

1971b   "César-François Cassini de Thury (Cassini III)." *DSB:* vol. 3, 107–109.

Teall, John L.
1959   "The Grain Supply of the Byzantine Empire, 330–1025." *Dumbarton Oaks Papers,* no. 13: 86–139.

Temkin, Owsei
1962   "Byzantine Medicine: Tradition and Empiricism." *Dumbarton Oaks Papers,* no. 16: 95–115.

Tennent, James Emerson
   1977   *Ceylon.* 3 vols. Dehiwala, Sri Lanka: Tisara Prakasakayo.
Thomas, Keith
   1971   *Religion and the Decline of Magic.* New York: Charles Scribner's
           Sons.
Thompson, J. E. S.
   1974   "Maya Astronomy." *Philosophical Transactions of the Royal Society
           of London* 276 (ser. A): 83–98.
Thorndike, Lynn
   1926–1927   "The Blight of Pestilence on Early Modern Civilization." *Ameri-
           can Historical Review* 32:455–74.
Thucydides
   1900   *Thucydides.* Vol. 1. Oxford: Oxford University Press.
Toulmin, Stephen, and June Goodfield
   1965   *The Fabric of the Heavens.* New York: Harper & Row.
Trigger, Bruce G.
   1980   *Gordon Childe.* New York: Columbia University Press.
Trigger, Bruce G., B. J. Kemp, D. O'Connor, and A. B. Lloyd
   1983   *Ancient Egypt: A Social History.* Cambridge: Cambridge University
           Press.
Turner, Billie Lee, II
   1974   "Prehistoric Intensive Agriculture in the Mayan Lowlands." *Science*
           185:118–24.
   1978   "The Development and Demise of the Swidden Thesis of Maya Agri-
           culture." In Harrison and Turner, 1978: 13–22.
Turner, Billie Lee, II, and Peter D. Harrison
   1983   *Pulltrouser Swamp.* Austin: University of Texas Press.
Udovitch, A. L., ed.
   1981   *The Islamic Middle East, 700–1900: Studies in Economic and Social
           History.* Princeton: Darwin Press.
van Andel, Tjeed H., and Curtis Runnels
   1987   *Beyond the Acropolis.* Stanford: Stanford University Press.
van der Waerden, B. L.
   1978   "Mathematics and Astronomy in Mesopotamia." *DSB:* vol. 15, 667–
           80.
von Hagen, Victor Wolfgang
   1961   *The Sun Kingdoms of the Americas.* Cleveland: World Publishing.
Vööbis, Arthur
   1965   *History of the School of Nisibis.* Louvain, Belgium: Secretariat du
           CorpusSCO.
Walker, Richard L.
   1956–1957   "Hydraulic Society." *Yale Review,* n.s. 46:601–603.
Watson, Andrew M.
   1974   "The Arab Agricultural Revolution and Its Diffusion, 700–1100." *The
           Journal of Economic History* 34, no. 1 (March): 8–35.
   1983   *Agricultural Innovation in the Early Islamic World.* Cambridge:
           Cambridge University Press.

Weber, Max
1976   *The Agrarian Sociology of Ancient Civilizations.* London: NLB.
Weisheipl, J. A.
1984   "Ockham and the Mertonians." In Catto, 1984: 607–58.
Wenke, Robert J.
1980   *Patterns in Prehistory.* New York: Oxford University Press.
Westfall, Richard S.
1975   "The Role of Alchemy in Newton's Career." In Bonelli and Shea, 1975: 189–232.
1980   *Never at Rest.* Cambridge: Cambridge University Press.
White, Lynn, Jr.
1962a   "The Agricultural Revolution of the Early Middle Ages." In White, 1962b: 39–78.
1962b   *Medieval Technology and Social Change.* Oxford: Oxford University Press.
Widtsoe, John A.
1914   *Principles of Irrigation Practice.* New York: Macmillan.
1987   *Evidences and Reconciliations.* Salt Lake City: Bookcraft.
Wiener, Philip, and Aaron Noland, eds.
1957   *Roots of Scientific Thought.* New York: Basic Books.
Wilford, John Noble
1989   "Guatemalan Find Alters Theories on Mayans." *New York Times,* November 14: C15.
Willey, Gordon R.
1966   *An Introduction to American Archaeology.* Vol. 1. Englewood Cliffs, N.J.: Prentice-Hall.
Williamson, Ray A.
1981a   "North America: A Multiplicity of Astronomies." In Williamson, 1981b: 61–80.
Williamson, Ray A., ed.
1981b   *Archaeoastronomy in the Americas.* Los Altos, Calif.: Ballena Press.
Williamson, Ray A., H. J. Fisher, and D. O'Flynn
1977   "Anasazi Solar Observatories." In Aveni, 1977: 203–17.
Willmot, W. E., ed.
1972   *Economic Organization in Chinese Society.* Stanford: Stanford University Press.
Wittfogel, Karl
1938   "Die Theorie der orientalischen Gesellschaft." *Zeitschrift für Sozialforschung* 7:90–122.
1981a   "The Theory of Oriental Society." In Bailey and Llobera, 1981: 141–57.
1981b   *Oriental Despotism.* New York: Random House.
Worster, Donald
1977   *Nature's Economy: A History of Ecological Ideas.* Cambridge: Cambridge University Press.
1985   *Rivers of Empire.* New York: Pantheon Books.

Worster, Donald, ed.

1988   *The Ends of the Earth: Perspectives on Modern Environmental History.* Cambridge: Cambridge University Press.

Yabuuti, Kiyosi

1973   "Chinese Astronomy: Development and Limiting Factors." In Nakayama and Sivin, 1973: 91–103.

Yates, Frances

1967   "The Hermetic Tradition in Renaissance Science." In Singleton, 1967: 255–74.

Zunz, Olivier, ed.

1985   *Reliving the Past.* Chapel Hill: University of North Carolina Press.